Science,
Technology
and
Innovation

Science, Technology and Innovation

A Research Bibliography

Compiled by Felicity Henwood
with Graham Thomas

Science Policy Research Unit

St. Martin's Press   New York

016.6
H528

©Science Policy Research Unit, 1984
All rights reserved. For information, write:
St. Martin's Press, Inc., 175 Fifth Avenue, New York, NY 10010
Printed in Great Britain
First published in the United States of America in 1984

ISBN 0-312-70281-7

Library of Congress Cataloging in Publication Data

Henwood, Felicity, 1957-
  Science, technology, and innovation.

  Includes index.
    1. Technological innovation — Bibliography.
I. Thomas, Graham, 1951-   II. University of Sussex.
Science Policy Research Unit.   III. Title.
Z7914.T247H46   1984   [T173.8]   016.6   83-40179
ISBN 0-312-70281-7

ACKNOWLEDGEMENTS

This bibliography has been compiled with the assistance and encouragement of many people at the Science Policy Research Unit. There are, however, several people who deserve specific mention.

In particular, I would like to thank my colleague Graham Thomas who, with the assistance of Chris Wimlett of the University of Sussex Computing Centre, prepared the computer programmes and procedures for the sorting and classification of references and who overcame the many technical problems that arose during the preparation of the bibliography. He also provided valuable criticism and support throughout.

I should also like to thank Jane Hamed, Vivien Johnson and Glenice Russell of the Science Policy Research Unit library who helped to select the original collection of references from which the final selection was made. They all gave me much of their time, advice and assistance.

The typing of references onto the computer was almost solely undertaken by Sue Plummer. Her contribution to the bibliography was therefore invaluable, as were her criticisms and comments regarding specific references.

I would also like to thank Chris Freeman and Keith Pavitt whose knowledge of the literature in this field enabled them to make very valuable contributions. Chris Freeman suggested an initial classification system, which, with a few modifications, is reflected in the final layout. Keith Pavitt read drafts and made suggestions for additional references and was also responsible for part of the introduction. My thanks to them both.

Finally, I should like to acknowledge Wheatsheaf Books Ltd. who commissioned this piece of work and the Leverhulme Trust who provided financial support without which the bibliography could not have been completed.

While all the above-mentioned people have helped make the bibliography possible, the responsibility for the final selection and arrangement of references remains mine.

Felicity Henwood

May 1983

CONTENTS

Introduction.   ix

Chapter 1. Measurement ot Science and Technology.   1

(a) R & D statistics   1
(b) "Scientometrics", publication counts, citation
    analysis   4
(c) Inventions and patents statistics   5
(d) Innovation statistics   7
(e) Technology transfer statistics   7
(f) Other statistics of science & technology   8
(g) Problems of measuring science & technology   10

Chapter 2. Inventions and Patents.   12

(a) Case studies of inventions   12
(b) The invention process   13
(c) Legal, economic and social analysis of patents
    and inventions   14

Chapter 3. Innovations.   20

(a) Innovations and economic development: case
    studies, theory   20
(b) Diffusion of innovations: firms and countries   40
(c) Specific new technologies   51
(d) Social innovation and technical change   58

Chapter 4. Economics of Technical Change.   59

(a) Macro-level analysis: technical change and
    economic growth   59
(b) Firm behaviour, competition, market and
    industrial structure   80
(c) International trade, technical change and
    innovation   101
(d) Technical change and regional economic
    development   113

Chapter 5. Social and Political Aspects of
           Technical Change and Innovation.   116

(a) The decision-making environment ot innovation   116

(b) Ecological environment, technical change and
    innovation   119
(c) Social effects of technical change and
    innovation   121

Chapter 6. Management of Technical Change and
           Innovation.   124

(a) Organisation and management of R&D and
    scientific laboratories   124
(b) Project evaluation in R&D   157
(c) Technology assessment   162
(d) Other aspects of management   163

Chapter 7. Technical Change, Innovation and Government. 176

(a) The effect of government policies on innovation   176
(b) National policies for technical change and
    innovation   189
(c) Other policies on technical change and innovation
    (OECD, EEC policy etc.)   205

Chapter 8. Technical Change and Work.   217

(a) The effects of technical change on job content   217
(b) Technical change, employment and unemployment   225

Chapter 9. Other Bibliographies on Technical Change
           and Innovation.   233

List of Relevant Journals and Periodicals.   239

Keyword Index.   240

INTRODUCTION

This bibliography concentrates on the subject of technical innovation which we define to include all processes associated with the generation, adoption and diffusion of new and better products, production processes and systems. Technical innovation thus defined has been an essential feature of economic growth and structural change during the past 200 years.

Since the second world war, academic social scientists and practitioners in industry, government and elsewhere have been made more keenly aware of the importance of technical innovation. So called "growth accounting" exercises by economists in the 1950s and 1960s showed that a large proportion of economic growth could be explained statistically only through a "residual" that included technical innovation. Similarly in the 1960s and 1970s patterns of comparative advantage of the industrially advanced countries in export markets were best explained by theories that made technical innovation a central feature of the competitive process. Now in the economically depressed and uncertain 1980s, more attention is given to the effects of technical innovation on skills, employment and opportunities for investment.

Whatever the particular concern at any one time, a better understanding of the nature and determinants of technical innovation is essential for the design and improvement of policies to guide and stimulate innovation in the appropriate direction and at the appropriate speed. However, such better understanding does not come easily, for at least three reasons.

- First, there is the unsatisfactory and incomplete treatment of technical innovation in many of the social sciences. Whilst "technical change" is recognised as a major feature of economic, social and political processes in modern society, it is often made exogenous to models and analysis and it is assumed to be homogenous in its effects.

- Second, there are severe problems of empirical measurement. Until recently, social scientists and policy analysts have had to rely either on very indirect measures of technical change (for example, "total factor productivity" or "non-price factors" in economic analysis) or on the time consuming collection of primary data from micro and case studies. Recent improvements have, however, been made in the collection of comprehensive data that measure technical innovation more directly.

- Third, there is the problem of the dispersion or what Nelson and Winter have called the "balkanisation" of studies of technical innovation.

Some important contributions towards a better understanding of the innovation process have been made recently by practitioners and journalists and also in and around the edges of a number of disciplines such as history, economics, sociology and the management and political sciences. As yet, however, there is not a large and clearly defined intellectual community primarily concerned with analysing problems of technical innovation. Communication is, therefore, very important.

Under these circumstances, the task of treating technical innovation as the primary subject of analysis is a difficult but challenging one. The Science Policy Research Unit has, from its beginning, been concerned to understsnd the nature, determinants and consequences of technical innovation, specifically. As a consequence of its research programmes and projects, it has built up a library collection that is probably unique in its primary focus and in the diversity of its coverage of disciplinary and published sources.

The following bibliography is based on the SPRU library collection and reflects these characteristics. Although the main focus is technical innovation, its content is eclectic and multi-disciplinary. It includes a broad range of references: from those that examine the role of basic science in technical innovation to those that discuss the social implications of specific new technologies. As such, it should be of use to practitioners in industry and elsewhere, to policy makers in industry and government, and to scholars from a variety of disciplines both in the natural and the social sciences.

Before discussing the layout of the bibliography and the arrangement of the references, there are several limitations and biases that should be made explicit. The bibliography undoubtedly reflects the specific interest of SPRU researchers and in this sense, at least, cannot claim to be comprehensive.

The selection of references concentrates on literature from European or OECD countries and largely excludes material emanating from or pertaining to developing or newly industrialising countries. It was felt that the inclusion of such refrences would make this bibliography too large and that there was sufficient literature on developing countries to warrant a separate bibliography. The exclusion of Soviet and other Eastern European literature reflects, again, the Unit's concentration on OECD countries and thus the small amount of material on other countries that is available to us. Non-English language references have not been included in the bibliography as the library holds a fairly random collection reflecting, in large part, languages spoken by SPRU researchers.

We have endeavoured to make the bibliography as comprehensive as possible. Those relevant titles not listed may be traced through the bibliographies section (Chapter 9), which contains a list of relevant bibliographies, literature reviews and "core papers" to which users should refer for further reading.

The bibliography contains over 2,000 references which are arranged into 9 sections. It is intended that the contents list provide users with a guide to the material and to specific references pertaining to their subject. However, users may also wish to refer to the keyword index, where individual reference numbers are listed alongside the keywords that best describe the content of those references.

A list of journals which regularly contain articles directly related to the subject of technical innovation can be found on p.239. It would have been impractical to list, individually, all the many hundreds of articles contained in these journals and therefore articles from them are included only where they were thought to be particularly interesting and relevant. Other articles that appear in this bibliography come from journals that do not regularly carry articles on technical innovation; they have been selected because of their particular relevance to the subject area.

Finally, it should be noted that this bibliography was prepared on computer and a database of references on technical innovation has been established with the intention of keeping it up to date. SPRU library would therefore welcome notice of any ommissions that might usefully be added to this collection.

Felicity Henwood
Keith Pavitt

May 1983

CHAPTER 1. <u>Measurement Of Science & Technology</u>

1.a R & D Statistics

1.001 ARNOW, KATHRYN S.
'Inaicators of price and cost change in research and development inputs'. In: <u>American Statistical Association</u>, Proceedings of the Business and Economic Statistics Section. Washington D.C.: American Statistical Association, 1966.

1.002 CANADA − Department of Industry
<u>Statistical data on industrial research and development in Canada.</u> Presented by J.L. Orr. Ottawa, Queen's Printer, 1967.

1.003 CANADA − Dominion Bureau ot Statistics
<u>Statistics ot scientific research and development in Canada.</u> Ottawa, Queen's Printer, 1966.

1.004 CANADA − Dominion Bureau ot Statistics
<u>Inaustrial research and development expenditure in Canada. 1967.</u> Ottawa, Queen's Printer, 1970.

1.005 FISHER, W. Halder
<u>Study of availability and quality of national statistics on R & D in selected OECD countries.</u> Final report and summary report. Columbus, Ohio: Battelle Columbus Laboratories, 1974. 114 leaves + 11 leaves.

1.006 FREEMAN, C.
<u>Measurement of output ot research and experimental development.</u> A review paper. Paris: Unesco, 1969.

1.007 GREAT BRITAIN − Business Statistics Office
<u>Business Monitor: industrial research and development expenditure and employment. 1978.</u> London: HMSO, 1980. (MO14).

1.008 LINK, Albert N.
'Rates or induced technology from investments in research ana development'. <u>Southern Economic Journal</u>, Vol. 45, No. 2, October 1978, p. 370−379.

1.009   ORGANISATION FOR ECONOMIC CO-OPERATION AND
         DEVELOPMENT
Patterns or resources devoted to research and experimental
development in the OECD area, 1963-71. Paris: OECD, 1975.
113p.

1.010   ORGANISATION FOR ECONOMIC CO-OPERATION AND
         DEVELOPMENT
Current research on indicators of R & D outputs and their
economic effects. By C. Kitti. Second workshop on the
measurement of R & D Output, 5th and 6th December, 1979.
Paris, OECD, 1979. 14 leaves. Mimeo.

1.011   ORGANISATION FOR ECONOMIC CO-OPERATION AND
         DEVELOPMENT
A proposed market-oriented continuum of R & D inputs,
outputs and innovation. By Kathryn S. Arnow. Second workshop
on the measurement ot R & D output, 5th and 6th December,
1979. Paris, OECD, 1979. 13 leaves. Mimeo.

1.012   ORGANISATION FOR ECONOMIC CO-OPERATION AND
         DEVELOPMENT
Science and technology indicators. Basic statistical
series - volume A - the objectives of government R & D
funding, 1969-81. Paris: OECD, 1981. 97p. Mimeo.

1.013   ORGANISATION FOR ECONOMIC CO-OPERATION AND
         DEVELOPMENT
Science and technology indicators, basic statistical series
- volume B: gross national expenditure on R & D (GERD)
1963-1979. Paris: OECD, 1982. 290p. Mimeo.

1.014 SIRILLI, Giorgio
'Linking R & D input and output indicators'. Quaderno, 6-7
September 1979, pp.53-84.

1.015 UNESCO
Statistics on research and experimental development
activities, 1967. Conference of Ministers of the European
Member States responsible for science policy. Paris, 22-27
June 1970. Paris: Unesco, 1970. 64p.

1.016 UNESCO
Recent statistical data on European research and experimental development activities (addendum). Meeting of experts
on science policy selected by the European member states,
Budapest, 4-7 July 1972. Unesco, 1972. 55 leaves. Mimeo.

1.017 UNESCO
The quantitative measurement of scientific and technological

activities related to research and experimental development.
By Jean-Claud Bochet. Paris: Unesco, 1975. 53p. (Current
Surveys and research in statistics).

1.018 UNESCO
Estimation of human and financial resources devoted to R & D
at the world and regional level. Paris: Unesco, 1979. 73p.
Mimeo.

1.019 UNESCO
Statistics on research and experimental development in the
European and North American Region. Paris: UNESCO, 1979.
114p.

1.020 UNESCO
Participation of women in R & D - a statistical study.
Paris: UNESCO, Office of Statistics, 1980. 53p. Mimeo.
(Current Studies and Research in Statistics).

1.021 UNITED STATES - Congress House Committee on Science
         and Astronautics: Subcommittee on science, research
         and development
Government and science. No.2: fiscal trends in federal
research and development. Washington, USGPO, 1964. 88th
Congress, 2nd session.

1.022 UNITED STATES - National Science Foundation
Basic research, applied research and development in
industry, 1962 - . Washington, USGPO.

1.023 UNITED STATES - National Science Foundation
Expenditure for scientific and engineering activities at
universities and colleges. Fiscal years 1958 - .
Washington, D.C.: USGPO.

1.024 UNITED STATES - National Science Foundation
Federal funds for research, development and other
scientific activities. Fiscal years 1962 - . Washington,
USGPO.

1.025 UNITED STATES - National Science Foundation
Research and Development in Industry - funds, scientists
and engineers, 1953 - . Washington, D.C., N.S.F.

1.026 UNITED STATES - National Science Foundation
R & D activities in state government agencies. Fiscal
years 1964 - . Washington, D.C.: USGPO.

1.027 UNITED STATES - National Science Foundation
Research expenditures of foundations and other non-profit

<u>institutions, 1953 - </u>. Washington, D.C.: USGPO.

1.028 UNITED STATES – National Science Foundation
<u>Government-university relationships in federally sponsored</u>
<u>scientific research and development.</u>   Washington,   USGPO,
1958.

1.029 UNITED STATES – National Science Foundation
'Trends in funds and personnel for research and development,
1953–61'. <u>Reviews of Data on Research and Development</u> No.33,
April 1962.

1.030 UNITED STATES – National Science Foundation
'Research funds used in the   nation's   scientific   endeavor,
1963'. <u>Reviews of Data on Science Resources</u>, 1(4), May 1965.

1.031 UNITED STATES – National Science Foundation
<u>R & D activities of independent non-profit institutions:</u>
<u>1973.</u> Washington: USGPO, 1975. 72p.

1.032 UNITED STATES – National Science Foundation
<u>National patterns of R & D resources: funds and manpower in</u>
<u>the United States, 1953-1976.</u>   Washington,   D.C.:   USGPO,
1976. 32p.

1.b "Scientometrics", Publication Counts, Citation Analysis

1.033 BROOKS, Harvey
'Science indicators and science policy'. <u>Scientometrics</u>,
Vol.2, No.5–6, October 1980, p.331–337.

1.034 CARPENTER, Mark
<u>Development of refined indicators of technology innovation</u>
<u>using examiners' citations in the patent file.</u> Cherry Hill,
N.J.: Computer Horizons, Inc., 1981. 26 leaves. Mimeo.

1.035 CARPENTER, Mark P. and NARIN, Francis
'The adequacy of the Science Citation Index (SCI) as an
indicator   of   international   scientific   activity'. <u>The</u>
<u>Journal of the American Society for Information Science</u>,
November 1982, p.430–439.

1.036 CARPENTER, Mark P., NARIN, Francis and WOOLF,
       Patricia
'Citation rates to technologically important patents'.
<u>World Patent Information</u>, Vol.3, No.4, 1981, p160–163.

1.037 GREAT BRITAIN – Department of Education and Science
<u>An attempt to quantify the economic benefits of scientific</u>
<u>research.</u> By I.C.R. Byatt and A.V. Cohen. London, HMSO,
1969. Science Policy Studies, 4.

1.038 GREAT BRITAIN – Department of Education and Science
<u>Research environment and performance in British university</u>
<u>chemistry.</u> London: HMSO, 1973. 53p. (D.E.S. science policy
studies, no. 6).

1.039 McALLISTER, Paul R. and WAGNER, Deborah Ann
'Relationship between R & D expenditure and publication
output for U.S. colleges and universities'. <u>Research in</u>
<u>Higher Education,</u> Vol. 15, No. 1, 1981, p. 3-30.

1.040 UNITED STATES – National Science Board
<u>Science Indicators, 1976.</u> Washington, D.C.: N.A.B., 1977.
304p.

1.c Inventions & Patents Statistics

1.041 BASBERG, Bjorn L.
<u>Technological change in the Norwegian whaling industry: a</u>
<u>case-study in the use of patent statistics as a technology</u>
<u>indicator.</u> Bergen-Sanviken: Norwegian School of Economics
and Business Adminisration. 19p. Mimeo.

1.042 BOSWORTH, Derek L.
<u>Statistics of technology (invention and innovation).</u>   (No
Imprint, no date). 127p. Mimeo. (Reviews of UK Statistical
Sources; edited by W F Maunder, No 28 – provisional).

1.043 BOSWORTH, Derek L.
<u>Changes in the quality of inventive output and patent based</u>
<u>indices of technological change.</u> Coventry: Centre   for
Industrial Economic and Business Research, University of
Warwick, 1972. 13 leaves.

1.044 CARPENTER, Mark
<u>Ten year trends in U.S. technological dependence on foreign</u>
<u>science and technology as indicated by referencing from U.S.</u>
<u>patents.</u> Cherry Hill, N.J.: Computer Horizons Inc., 1981.
40 leaves and Appendices. Mimeo.

1.045 UNITED STATES – Department of Commerce, Patent and
        Trademark Office
<u>Patent profiles: solar energy.</u> Washington D.C.:   USGPO,

January 1980. 190p.

1.046   UNITED STATES - Office of Technology Assessment and
        Forecast
U.S. patenting in 55 standard industrial classification
product fields (1963-1979). A special report prepared for
the Science Indicators Unit, National Science Foundation. No
imprint. No pagination. (Indicators of Patent Output of U.S.
Industry, IV).

1.047   UNITED STATES - Office of Technology Assessment and
        Forecast
Patent activity profile: all technologies. Special report.
Washington   D.C.:   Office   of   Technology   Assessment and
Forecast. No date. No pagination. Mimeo.

1.048   UNITED STATES - Office of Technology Assessment and
        Forecast
Energy patenting (1963-1979). U.S. Patent   and   Trademark
Office.   Washington   D.C.:   NSF,   1980.   3 leaves and
microfiche.

1.049   UNITED STATES - Office of Technology Assessment and
        Forecast
Patent profiles - Microelectronics I. Washington,   D.C.:
USGPO, 1981. 257p.

1.050   UNITED STATES - Office of Technology Assessment and
        Forecasting
U.S. patent activity in 39 standard industrial classification categories, 1963 - 1976E (considering original - or
- and cross reference - XR - patent classifications). Report
prepared for the Science Policy Research Unit by the Office
of Technology Assessment and Forecast and U.S. Patent and
Trademark Office. No imprint. Not paginated.

1.051   UNITED STATES - Office of Technology Assessment and
        Forecasting
Company listing and patent activity profile (all techno-
logies). Report prepared for the Science Policy Research
Unit by the Office of Technology Assessment and Forecast and
U.S. Patent and Trademark Office. No imprint. 321p.

1.d   Innovation Statistics

1.052   AUSTRALIAN BUREAU OF STATISTICS
Technological change in private non-farm enterprises in
Australia for the three years ending 30 June 1979 - detailed
report. Canberra: Australian Bureau of Statistics, 1980. 37
leaves. Mimeo.

1.053   CANADA - Statistics Canada
Selected statistics on technological innovation in industry.
Ottawa: Information Canada, 1975. 46p. In English and
French.

1.054   KLEIMAN, Herbert S.
Indicators of the output of new technological products from
industry. Ohio: Battelle Columbus Labs, 1975. Microfiche.
PB-293 0/2.

1.055   ORGANISATION FOR ECONOMIC CO-OPERATION AND
        DEVELOPMENT
L'observation et la mesure de l'innovation. By Andre Piat-
ier. Paris, OECD, 1979. 35 leaves. Mimeo.

1.056   ORGANISATION FOR ECONOMIC CO-OPERATION AND
        DEVELOPMENT
Workshop on patent and innovation statistics – held   at
Chateau de la Muette, Paris, 28,29 and 30 June 1982. Paris:
OECD, 1982. Assorted Papers. Mimeo.

1.057   TOWNSEND, J.F., HENWOOD, F., THOMAS, G., PAVITT, K.
        and WYATT, S.
Science and technology indicators for the U.K. - innovations in Britain since 1945. Brighton: SPRU, December
1981. 129p. (SPRU Occasional Paper No.16).

1.e   Technology Transfer Statistics

1.058   ORGANISATION FOR ECONOMIC CO-OPERATION AND
        DEVELOPMENT
Data concerning the balance of technological payments in
certain OECD member countries: statistical data and
methodological analysis. Paris: OECD, 1977. 110p. Mimeo.

1.059   ORGANISATION FOR ECONOMIC CO-OPERATION AND
        DEVELOPMENT
Workshop on the Technological Balance of Payments - held 14

and 15 December 1981. Paris: OECD, 1981. Assorted papers. Mimeo.

1.060 UNITED STATES - Department of Commerce
Alternative measurements of technology-intensive trade.
Prepared by Regina K. Kelly. A staff Economic Report. Springfield, Va.: U.S. Department of Commerce, 1976. 20p. Mimeo.

1.061 UNITED STATES - National Science Foundation
Indicators of international trends in technological innovation. Final report prepared by Stephen Feinman and William Fuentevilla. Springfield, Va.: National Technical Information Service, 1976. 93p plus appendices.

1.f Other Statistics Of Science & Technology

1.062 CANADA - Statistics Canada
Annual review of science statistics 1978. Ottawa: Statistics Canada, 1979. 122p.

1.063 EUROPEAN COUCIL OF CHEMICAL MANUFACTURERS' FEDERATIONS
Chemical industry basic economic statistics, 1979-1980.
Brussels: CEFIC, 1981. No pagination. Mimeo.

1.064 FREEMAN, C.
The measurement of scientific and technological activities: proposals for the collection of statistics on science and technology on an internationally uniform basis.   Paris, Unesco, 1969.

1.065 O'BRIEN, Ronan
Feasability study of S & T indicators for Irish manufacturing industry. Final report. Dublin: University College. Science Policy Research Centre, 1980. 71p.

1.066 ORGANISATION FOR ECONOMIC CO-OPERATION AND DEVELOPMENT
Science and technology in the new socio-economic context - data on growth trends of labour productivity. Paris: OECD, 1979. 38p. Mimeo.

1.067 PAVITT, Keith
'R & D, patenting and innovative activities: a statistical exploration'. Research Policy, Vol.11, No.1, February 1982, pp.33-51.

1.068 ROSENBERG, Nathan
An examination of international technology flows in Science
Indicators 1978. Stanford, California: Stanford University,
1980. 26 leaves. Mimeo.

1.069 UNESCO
Statistical Yearbook, 1967. Paris, 1968.

1.070 UNESCO
World summary of statistics on science and technology.
Results of experimental surveys of scientific and technical
manpower and expenditures for research and experimental
development.

1.071 UNESCO
Guide to the collection of statistics on science and
technology. Paris: Unesco, 1977. 110p.

1.072 UNESCO
Recommendation concerning the international standardization
of statistics on science and technology adopted by the
General Conference at its 20th session, Paris, 27 November
1978. Paris: UNESCO, 1978. 70p.

1.073 UNESCO
Statistics on science and technology - latest available
data. Paris: UNESCO, 1980. 157p. (Current Surveys and
Research in Statistics).

1.074 UNITED NATIONS
Statistical Yearbook, 1968. New York, 1969.

1.075 UNITED STATES - National Science Foundation
National patterns of science and technology resources 1980.
Washington D.C.: USGPO, 1980. 73p.

1.076 UNITED STATES - National Science Foundation
Papers commissioned as background for Science Indicators -
1980. Five volumes. Washington D.C.: N.S.F., 1980. Vol.1:
Indicators of international technology and trade flows.
Vol.2:   Indirect mechanisms of federal support of research
and development. Vol.3: Administrative responsibilities and
the   conduct   of academic basic research. Vol.4:   The
measurement of industrial innovation. Vol.5: The adequacy
of job opportunities for recent science and engineering
graduates.

1.077 WORLD INTELLECTUAL PROPERTY ORGANIZATION
Industrial property statistics 1977 in the form of summary
tables. Geneva: WIPO, 1979. 368p. (Publication B).

1.g Problems Of Measuring Science & Technology

1.078 DOMAR, Evsay D.
'On the measurement of technological change'. Economic Journal, December 1961, Vol.71(284), 709-729.

1.079 FLUECKIGER, Gerald E.
'Observation and measurement of technological change', Explorations in Economic History, Vol.9, No. 2, Winter 1971-2, p.p. 145-177.

1.080 GORDON, Gerald
'The problem of assessing scientific accomplishment: a potential solution'. IEEE Transactions, Vol. EM-10(4), December 1963, pp.192-196.

1.081 GORDON, Theordore J. and MUNSON, Thomas R.
A proposed convention for measuring the state of the art of products or processes. Glastonbury, Conn.: The Futures Group, 1981. 39 leaves. Mimeo.

1.082 ORGANISATION FOR ECONOMIC CO-OPERATION & DEVELOPMENT
The meaning of patent statistics. Second workshop on the Measurement of R & D Output, 5-6 December, 1979. Paris, OECD, 1979. 104p. Mimeo.

1.083 ORGANISATION FOR ECONOMIC CO-OPERATION AND
        DEVELOPMENT
The measurement of scientific and technical activities: proposed standard practice for surveys of research and development. OECD, 1967.

1.084 ORGANISATION FOR ECONOMIC CO-OPERATION AND
        DEVELOPMENT
The measurement of scientific and technical activities: proposed standard practice for surveys of research and experimental development. Paris, OECD, 1970. (Frascati manual).

1.085 ORGANISATION FOR ECONOMIC CO-OPERATION AND
        DEVELOPMENT
The measurement of scientific and technical activities.
Notes on a proposed standard practice for surveys of research in the social sciences and humanities, proposed addendum to the Frascati Manual (note to the Secretariat). Paris, OECD, 1971. 35p.

1.086 ORGANISATION FOR ECONOMIC CO-OPERATION AND
        DEVELOPMENT

The meaning of patent statistics and their use in science indicators. By Jennifer Sue Bond. 1978. Paper presented at the Second Workshop on the Measurement of R & D output at the OECD Headquarters, Paris, December 5 and 6, 1979. Paris, OECD, 1979. 7p. Mimeo.

1.087   ORGANISATION FOR ECONOMIC CO-OPERATION AND
         DEVELOPMENT
The development of indicators to measure the output of R & D: some preliminary results and plan for future work. Committee for Scientific and Technological Policy. Paris, OECD, 1979. 42p. Mimeo.

1.088   ORGANISATION FOR ECONOMIC CO-OPERATION AND
         DEVELOPMENT
The use of patent information as an indicator of technological output: the work of the Office of Technology Assessment and Forecast, United States Patent and Trademark Office. By W.S. Lawson. Second workshop on the measurement of R & D output, 5th and 6th December, 1979. Paris, OECD, 1979. 11p. + appendices. Mimeo.

1.089   ORGANISATION FOR ECONOMIC CO-OPERATION AND
         DEVELOPMENT
The measurement of scientific and technical activities: proposed standard practice for surveys of research and experimental development. Frascati Manual, 1980. Paris, OECD, 1981. 186p.

1.090   SAHAL, Devendra
'A theory of measurement of technological change'. The International Journal of Systems Science, Vol.8, No.6, 1977, pp.671-682.

1.091   UNITED STATES - General Accounting Office
Science indicators: improvements needed in design, construction, and interpretation. Report by the Controller General of the United States. Washington, D.C.: USGAO, 1979. 70p.

CHAPTER 2. Inventions And Patents

2.a Case Studies Of Inventions

2.001 BAKER, R.
New and improved... inventors and inventions that have
changed the modern world. London: British   Museum Publications Ltd., 1976. 168p.

2.002 CROWTHER, James Gerald
Discoveries and inventions of the 20th century. 6th ed. New
York, Dutton, 1966.

2.003 FAIRLEY, Peter
British inventions in the twentieth century. Norwich: Flet-
cher & Son, 1970.

2.004 FOX, M., CALGAR, M.L. and ROBERTSON, A.B.
Inventions from non-industrial sources. Complete report to
the Leverhulme Trust Fund of the two year study, June 1977-
June 1979, 2 vols. London: The   Polytechnic   of Central
London,   School   of   Management   Studies,   1980. Vol. 1 - 68
leaves. Vol. 2 - Appendices. Mimeo

2.005 LARSEN, Egon
A history of invention. London: J.M.Dent & Sons,1961.

2.006 LEHRBURGER, Egon
Ideas and inventions. London, Spring Books, 1960.

2.007 LYNN, Richard
The entrepreneur: eight case studies.   London: Allen   &
Unwin, 1974. 175p.

2.008 NATIONAL BUREAU OF ECONOMIC RESEARCH.
The rate and direction of inventive activity: economic and
social factors. National Bureau of Economic Research Special
Conference Series. Princeton University Press, 1962.

2.009 NEEDHAM, Joseph.
'China and the invention of the pound-lock', Transactions
of the Newcomen Society, Vol. 36, 1963-64, p. 85-107.

2.010 PARKER, R.C.
'Creativity: a case history'. Engineering, February 1975,
pp.126-130.

2.011 SCHMOOKLER, Jacob
'Inventors past and present'. Review of Economics and
Statistics 39(3), 1957, pp.321-333.

2.b The Invention Process

2.012 AHMAD, Syed
'On the theory of induced invention'. Economic Journal,
1966, 76(302), 344-357.

2.013 BARNES, Carl E.
'To promote invention'. International Science and Technology, December 1966, No.60, 67-73.

2.014 BERLE, Alf Keyser and DE CAMP, Lyon Sprague
Inventions, patents and their management. Princeton, N.J.,
Van Nostrand, 1959.

2.015 CALVERT, Robert (ed)
The encyclopedia of patent practice and invention
management. New York: Reinhold; London: Chapman & Hall,
1964. 860p.

2.016 ENCEL, Solomon and INGLIS, A.
'Patents, inventions and economic progress', The Economic
Record, December 1966, 572-588.

2.017 FISKE, Bradley Allen
Invention: the master-key to progress. New York, E.P. Dutton
& Co., 1921.

2.018 KINGSTON, William
Invention and monopoly. London, Woolwich Polytechnic, 1967.
Woolwich Economic Papers, 15.

2.019 PRICE, George R.
'How to speed up invention'. Fortune, November 1956, 54(5),
150-153.

2.020 RATTEE, I.D.
Discovery of invention? An inaugural lecture. Leeds: Leeds
University Press, 1964. 12p. Mimeo.

2.021 ROBERGE, R.
The timing, type and location of adaptive inventive
activity in the eastern Canadian pulp and paper industry:
1806-1940. Ann Arbor: University Microfilms. Published on
demand. 1977. 259p.

2.022 ROBERTSON, Andrew
'The inventive urge'. Industry Week, December 19th, 1969,
pp.28-29.

2.023 ROCKETT, John F.
'Licensing inventions'.   International Science and Technology, 1967, No.67, 33-42.

2.024 SCHMOOKLER, Jacob
'The level of inventive activity'. Review of Economics and
Statistics, 36(2), May 1954, pp.183-190.

2.025 SCHMOOKLER, Jacob
Patents, invention, and economic change: data and selected
essays. By Jacob Schmookler; edited by GRILICHES Zvi and
HURWICZ Leonid. Cambridge, Mass.: Harvard U.P., 1972. 292p.

2.026 THRING, Meredith
'A workshop for inventions'. New Scientist, 42(653), 12 June
1969, pp.576-577.

2.c Legal, Economic & Social Analysis Of Patents &
     Inventions

2.027 ALDERSON, W. (ed. with others)
Patents and progress: the sources and impact of advancing
technology. Homewood, III., Richard D. Irwin, 1965.

2.028 ALDOUS, GUY and others
Terrell on the law of patents. 11th ed. London, Sweet &
Maxwell, 1965.

2.029 BEHRMAN, J.N.
'Licensing abroad under patents, trademarks, and knowhow by
U.S. companies: report of a survey of problems and
practices and their relation to the American patent and
trademark systems'. Patent, Trademark and Copyright Journal
of Research and Education, June 1958, 2, 181-277.

2.030 BOEHM, Klaus
The British patent system - 1. Administration.   Cambridge,

U.P., 1967.

2.031 BRINK, Richard E. and others
An Outline of United States Patent Law. New York, Inter-
science, 1959.

2.032 CANADA - Economic Council of Canada
Report on intellectual and industrial property.   Ottawa:
Queen's Printer, 1971. 236p.

2.033 CARPENTER, Mark P., COOPER, Marti and NARIN, Francis
'Linkage between basic research literature and patents'.
Research Management, Vol. XXIII, No.2, March 1980, P.30-35.

2.034 COOPER, Charles, FREEMAN, C. and SERCOVICH, F.
The British patent system in relation to the international
patent system and developing countries. SPRU, 1973. 133p.
Mimeo.

2.035 DE CAMP, L. Sprague
'An early patent law proposal?' Technology and Culture,
Summer 1964, Vol.5(3), 377-379.

2.036 DOUGLAS, R.D.
'Commercial implications of patents of the pharmaceutical
industry'. Chemistry and Industry, 2 June 1973, p. 506-7.

2.037 DUCKIES, Robert A.
Ideas, inventions, and patents: how to develop and protect
them. New York, Wiley, 1957.

2.038 EATON, William W.
'Patent problem: who owns the rights?' Harvard Business
Review, July-August, 1967, Vol.45, 101-110.

2.039 FEDERICO, P.J.
'Historical patent statistics, 1791-1961'. Journal of the
Patent Office Society, February 1964, Vol.46(2), 89-171.

2.040 FERGUSSON, J.D.
'International developments in patent legislation'.
Chemistry and Industry, 2nd June 1973, p. 504-505.

2.041 FOREMAN, Howard I.
Patents: their ownership and administration by the United
States government. New York, Central Book Co., 1957.

2.042 GABAY, Mayer
The patent system and technological development in Latin
America. Washington, D.C.: Organization of American States,

1971. 9p.

2.043 GOMME, Allan A.
Patents ot invention: origin and growth of the patent
system in Britain. London, Longmans, 1946.

2.044 GREAT BRITAIN - Department of Trade
Patent law reform: a consultative document. London: HMSO,
1975. 46p.

2.045 GREAT BRITAIN - Patent Office
Patents: a source of technical information. London: HMSO,
1979. 36p.

2.046 GREVINK, H. and KRONZ, H.
Evolution of patent filing activities in the EEC: a contribution to the study and assessment of the technological
trends developing in the EEC from 1969 - 1975, based on a
statistical analysis of patents. Commission of the European
Communities, Information Management, 1979. 368 leaves. (EUR
6574 e).

2.047 GREVINK, H. and KRONZ, H.
Evolution of patent filing activities in the USA: a
contribution to the study and assessment of technological
trends in the USA from 1969 to 1974, based on a statistical
analysis of patents. Commission of the European Communities,
1980. 132 leaves. (EUR 6575e).

2.048 HAFSTAD, Lawrence R.
'Lay comments on the proposed patent   law'.   Research
Management, 1969, Vol.12(2), 107-121.

2.049 HAWKINS, Lawrence Ashley
'Does patent consciousness interfere with   cooperation
between industrial and university research laboratories?'
Science, 28 March 1947, Vol.105, 326-327.

2.050 HOLMAN, Mary A.
'The utilization of government-owned patented innovations'.
Patent, Trademark and Copyright Journal of Research and
Education. Summer-Fall 1963, Vol. 7, 109-161, 321-275.

2.051 HUMMERSTONE, Robert G.
'How the patent system mousetraps inventors'. Fortune, May
1973, p. 262-274.

2.052 JEWETT, F B.
'Modern research organizations and the American patent
system. Mechanical Engineering, June 1932, Vol. 54, 394-398.

2.053 JONES, Stacy V.
You ought to patent that. New York, Dial Press. 1972.

2.054 KAHN, Alfred E.
'Fundamental deficiencies of the American patent law'.
American Economic Review, 1940, Vol. 30, 478.

2.055 KINGSTON, William
'Future of British patents'. New Scientist, 6 August, 1970,
Vol. 47(713), 291-292.

2.056 KITTI, Carole and TROZZO, Charles L.
The effects of patent and antitrust laws, regulations, and
practices on innovation. Virginia: Institute for Defense
Analyses Program Analysis Division, 1976. 3 Vol. IDA Log no.
HQ 76-18304. Vol. I: a state of the art review. Vol. II:
Executive Summary. Vol. III: Annotated bibliography
additionally authored by John Driffill and Mary Summerfield.

2.057 KRONSTEIN, Heinrich and TILL, Irene
'A reevaluation of the International Patent Convention'.
Law and Contemporary Problems 1947, Vol. 12, 765-781.

2.058 MANDEVILLE, T.D., LAMBERTON, D.M. and BISHOP, E.J.
Economic effects of the Australian patent system. Canberra:
Australian Government Publishing Service, 1982. 228p.

2.059 MANSFIELD, Edwin, SCHWATZ, Mark and WAGNER, Samuel
'Imitation costs and patents: an empirical study'. The
Economic Journal, Vol.91, December 1981, pp.907-918.

2.060 MARSTRAND, Pauline
New developments in biotechnology and their relevance to
environmental policy. An analysis of worldwide patents
registered 1977 - June 1980. A preparatory study carried out
on behalf of the Science Policy Research Unit, University of
Sussex. Brighton: SPRU, September 1980. 33 leaves. Mimeo.

2.061 MUELLER, Dennis C.
'Patents, research and development, and the measurement of
inventive activity', Journal of Industrial Economics, Vol.
15, No. 1, Nov. 1966, p. 26-37.

2.062 OPPENLANDER, Karl Heinrich
'Patent policies and technical progress in the Federal
Republic of Germany'. I.I.C., Vol.8, No.2, 1977, pp.97-122.

2.063 PAVITT, Keith
'Using patent statistics in "Science Indicators":
possibilities and problems' in: Organisation for Economic

Co-operation and Development, The meaning of patent statistics. Second Workshop on the Measurement of R & D Output, 5-6 December, 1979. Paris, OECD, 1979, pp.63-104.

2.064 PENNINGTON, Robert R. (ed)
European patents at the crossroads. Papers delivered at a conference held in Munich, January 1976. London: Oyez Publishing, 1976. 251p.

2.065 PENROSE, Edith T.
The economics of the international patent system. Baltimore, John Hopkins Press, 1951.

2.066 PRESTON, Lee E.
'Patent rights under Federal R & D contracts'. 1963. In: Harvard Business Review. R & D management series... Vol.1.

2.067 REIK, Richard
'Compulsory licensing of patents'. American Economic Review, 1946, 36(5), 813-832.

2.068 SCHERER, F.M.
The propensity to patent. Chicago: Northwestern University, (no date), 38 leaves. Mimeo. Added to stock, October 1981.

2.069 SCHERER, F.M.
The economic effects of mandatory patent licensing. Paper presented at Germantown, Maryland in a public meeting of the U.S. Energy Research and Development Administration, January 12, 1976. Evanston, Illinois: Department of Economics, Northwestern University, 1976. 134p. Mimeo.

2.070 SOETE, Luc L.G. and WYATT, Sally M.E.
'The use of foreign patenting as an internationally comparable Science and Technology Output Indicator'. Scientometrics, Vol.5, 1983, pp.31-54.

2.071 SOLO, Robert A.
'Patent policy for government-sponsored research and development'. Idea, 10(2), 1966, pp.143-206.

2.072 STONEMAN, P.
Patenting activity: a re-evaluation of the influence of demand pressures. Coventry: University of Warwick Centre for Industrial Economic and Business Research, 1978. 26p. (Discussion paper no.84).

2.073 TAYLOR, C.T. and SILBERSTON, Z.A.
The economic impact of the patent system: a study of the British experience. Cambridge: C.U.P., 1973. 408p.

(University of Cambridge Department of Applied Economics, Monograph 23).

2.074 TOWALSKI, Z.
Enzyme technology patents as quantitative indicators of activity. Prepared for the FAST Programme, sub-programme Biosociety. Brussels: The Commission for the EEC, 1982. 123p. Mimeo. (Cl.4 Analysis of the Scientific Disciplines germane to Biotechnology).

2.075 UNITED STATES - National Science Foundation
The meaning of patent statistics. Washington D.C.: National Science Foundation, 1978. 104p. Mimeo.

2.076 WORLD INTELLECTUAL PROPERTY ORGANISATION
The role of patent information in research and development. Collection of lectures given at the Moscow Symposium organised by the World Intellectual Property Organisation (Moscow, October 7-11, 1974). Geneva: WIPO, 1975. 371p. Microfiche in back pocket.

CHAPTER 3. INNOVATIONS

3.a Innovations & Economic Development: Case Studies, Theory

3.001 ACHILLADELIS, Basil G.
'A study in technological history. Part 1 - The manufacture of Perlon (nylon 6) and caprolactam by IG FarbenIndustrie. Part 2 - The development of the BASF caprolactam process. Part 3 - The development of the Snia Viscosa caprolactam process'. Chemistry and Industry, December 5, 12 and 19, 1970.

3.002 ADAMS, Walter and DIRLAM, Joel B.
'Big steel, invention, and innovation'. Quarterly Journal of Economics, May 1966, 80(2), 167-189.

3.003 ADLER, Stephen F.
'Patents and innovation in the chemical industries'. Research Management, July 1980, p.30-35.

3.004 ALBU, Austen
Innovation in shipbuilding and marine engineering. Brighton: SPRU, December 1975. 41 leaves. Mimeo.

3.005 ALDER, Stephen
'Patents and innovation in the chemical industries'. Research Management, July 1980, p.30-35.

3.006 AMES, Edward
'Communications: research, invention, development and innovation'. American Economic Review, June 1961, Vol.51, No.3, p.370-381.

3.007 ANNALES DES MINES
Whole issue on innovation. February 1972, 110p.

3.008 ARTHUR D. LITTLE
Barriers to innovation in industry: opportunities for public policy changes, main report. Washington DC: Arthur D. Little., 1973. 162p. Prepared for the NSF contracts NSF C748 and C725.

3.009 BHANEJA, B. et al.
Technology transfer by Department of Communications: a
study of eight innovations. Ottawa: MOSSET, 1980. 48p.
Mimeo. (Background Paper No. 12).

3.010 BLACK, Ronald P.
Technological innovation in civilian public areas.   Falls
Church, Va., Analytic Services Inc., 1967.

3.011 BLAUG, Mark
'A survey of the theory of process-innovations'. Economica,
1963, 30(117), 13-32.

3.012 BMFT/NSF
BMFT/NSF Seminar on Technological Innovation. Bonn,   5-9
April 1976. Various papers.

3.013 BOLLINGER, Lynn, HOPE, Katherine and UTTERBACK,
       James M.
Technology and industrial innovation in Sweden: a review of
literature and hypotheses on new technology-based firms.
Cambridge, Mass.:   Center   for   Policy Alternatives, MIT/
Stockholm National Swedish Board for Technical Development,
1981. 33 leaves. Mimeo. (CPA/WP 81-07)

3.014 BOLTON, E.
'Development of Nylon'.   Industrial and Engineering Chemistry, January 1942, 34(1), 53-58.

3.015 BOSWORTH, Derek L.
The changing international sources of U.K. technology.
Coventry: Centre for Industrial Economic and Business
Research, University of Warwick, 1973. 31p.

3.016 BRAISFORD, S.F.
'Innovation in the evolution of the paper   industry'.
Paper Technology and Industry, Vol.21, July-August, 1980,
p.193, 194, 200 and 201.

3.017 BRESSON, Chris
Does Canada fail to innovate? Science Council of Canada,
1979. 22p. + annex. Mimeo.

3.018 BRODSKY, Neal H., KAUFMAN, Harold G. and TOOKER,
       John D.
University/industry cooperation: a preliminary analysis of
existing mechanisms and their relationship to the innovation
process. New York: New York University, Graduate School of
Public Administration, 1980. 97p.

3.019 BROWN, William H.
'Innovation in the machine tool industry'. <u>Quarterly Journal of Economics</u>, August 1957, Vol.71(3), 406-425.

3.020 BROZEN, Yale
'Invention, innovation, and imitation'. <u>American Economic Review</u>, 1951, Vol.41(2), 239-257.

3.021 BURCHARDT, A.
<u>Report on a literature search to elucidate the nature of innovation.</u> Department of Trade and Industry, 1970. 65p.

3.022 CAHN, R.
'Case histories of innovations'. <u>Nature,</u> 21 February 1970, Vol.225(5234), 693-695.

3.023 CANADA - Science Council of Canada
<u>Innovation in a cold climate: the dilemma of Canadian manufacturing.</u> Ottawa: Information Canada, 1971. 49p. (Report no. 15.)

3.024 CARTER, Charles F. and WILLIAMS, Bruce R.
<u>Investment in innovation.</u> London: MacDonald, 1971. 125p.

3.025 CHEUNG, Steven N.S.
<u>Report: contractual arrangements and the capturability of returns to innovation - report of a pilot.</u> Preliminary, March 1976. 29 leaves plus appendices. Mimeo.

3.026 COOK, L.G. and MORRISON, W. Adair
<u>Innovation case histories collected in connection with "The Origins of Innovation".</u> Mimeo. New York, General Research Laboratory, 1961.

3.027 COOK, L.G. and MORRISON, W. Adair
<u>The origins of innovation.</u> New York, General Electric Research Laboratory, 1961.

3.028 COOMBS, R., GIBBONS, M. and GARDINER, P.
'Innovation in the United Kingdom tractor industry'. <u>Omega,</u> Vol.9, No.3, 1981, p.255-265.

3.029 COOMBS, R.W., GARDINER, J.P., GIBBONS, M. and
       JOHNSTON, R.D.
<u>Incremental innovation and the U.K. tractor industry.</u>
Manchester: Department of Liberal Studies in Science, 1978. Not paginated. Programme of Policy Research in Engineering, Science and Technology.

3.030 COOPER, Robert G.

Project newprod: what makes a new product a winner?   An empirical study of successful versus unsuccessful industrial product innovation. Quebec: Quebec Industrial Innovation Centre, 1980. 192p.

3.031 CREIGHTON, J.W., JOLLY, J.A. and DENNING, J.A.
Enhancement of research and development output utilization efficiencies: linker concept methodology in the technology transfer process. Monterey: Naval Postgraduate School, 1972. 173p.

3.032 CROTT, R.
The EEC policy on isoglucose - completed. A case study of a rapidly growing field of biotechnology, its history, development and commercial opportunities. Prepared for the FAST   Programme,   sub-programme   Biosociety.   Brussels: Commission for the EEC, 1981. 113p. Mimeo. (Cl.4 Analysis of the Scientific Disciplines germane to Biosociety).

3.033 DAVIES, Duncan
Case studies of British technological achievements. London: Department of Industry, 1981. 59 items. Mimeo.

3.034 DOSI, Giovanni
'Technological paradigms and technological trajectories: a suggested interpretation of the determinants and directions of technical change'. Research Policy, Vol.11, No.3, June 1982, pp.147-162.

3.035 DU PONT
'Orlon: case history of a new fibre'. Fortune, October 1950, Vol.42(4), 107-132.

3.036 DUCHESNEAU, Thomas et al.
A study of innovation in manufacturing: determinants, processes, and methodological issues. Vol.1 - A panel study of the determinants of innovation in the U.S. footwear industry, Vol. 11 - Case studies of innovation decision making in the U.S. footwear industry. Orono, Maine: University of Orono, 1979. Vol.1 - 463p. Vol.11 - 167p. Mimeo.

3.037 EELS, F.R. and others
'Innovation and automation: a discussion based on case studies'.   Bulletin of the Oxford Institute of Statistics, 1959, Vol.21(3), 131-208.

3.038 ENCEL, Solomon
'Science, discovery and innovation: an Australian case history'.   International Social Science Journal,   1970,

Vol.22(1). Whole issue on the sociology of science.

3.039 ENOS, John Lawrence
Petroleum progress and profits: a history of process
innovation. Cambridge, Mass., MIT Press, 1962.

3.040 EUREKA
Innovative Engineering Design. Whole issue. Vol.1, No.1,
December 1980. 116p.

3.041 EUROPEAN SPACE AGENCY
Economic effects of space and other advanced technologies.
International conference held at the Palais de l'Europe,
Strasbourg, 28-30 April, 1980. Strasbourg: ESA, 1980. 270p.
(ESA SP-151).

3.042 FAGEN, M.D. (ed)
Impact: a compilation of Bell System innovations in science
and engineering which have helped create new industries and
new products. Bell Laboratories, 1971. 147p.

3.043 FELLNER, William
'Two propositions in the theory of induced innovations',
Economic Journal, Vol.71(282), 1961, p.p.305-308.

3.044 FERNILIUS, W. Conrad et al
Contribution of basic research to recent successful
industrial innovations. Final report. St. Louis, Missouri:
Industrial Research Institute, Research Corporation, 1979.
Microfiche. PB80-160179.

3.045 FIELDS, Gordon B.
The influence of insurance on technological development.
Washington: George Washington University, 1969. Program of
Policy Studies in Science and Technology, Staff Discussion
Paper, 405.

3.046 FISHER, Franklin M. et al
'The costs of automobile model changes since 1949',
Journal of Political Economy, Vol.70, No.5, October 1962,
p.p.433-451.

3.047 FISHLOCK, David.
'The state of British technology'. New Scientist, 1963,
No.331, p.p.623-633.

3.048 FREEMAN, C
'Innovation as an engine of growth: retrospect and
prospects', in Giersch, Herbert(ed.), Emerging
technologies: consequences for economic growth, structural

change and employment. Kiel Symposium 1981. Tubingen: J.C.B. Mohr (Paul Siebeck), 1982. pp.1-32.

3.049 FREEMAN, C.
'The determinants of innovation: market demand, technology, and the response to social problems'. Futures, Vol.11, No. 3, June 1979, p. 206-215.

3.050 FREEMAN, C., SOETE, L. and TOWNSEND, J.
Fluctuations in the numbers of product and process innovations, 1920-1980. Paper presented at OECD Workshop on Patent and Innovation Statistics, 28-30 June 1982. Brighton: SPRU, June 1982. 16p + appendices. Mimeo.

3.051 FRYERS, Gordon
'Profits, public opinion and innovation'. In: Science of Science Foundation, Innovation and profitability: three contributions to discussion. London, 1968.

3.052 FRYERS, Gordon
'Technological innovation and the growth of the economy'. Science Policy News, Vol. 1(1), July 1969, pp.18-21.

3.053 GEORGE, Frank (ed)
Science fact. Great Missenden, Bucks: Topaz Books, 1977. 540p.

3.054 GIBBONS, M. and JOHNSON, R.D.
'Relationship between science and technology'. Nature, Vol. 227(5254), 11 July 1970, pp.125-127.

3.055 GIBBONS, M. and JOHNSTON, R.D.
The interaction of science and technology. Final report of a study carried out for the Economic Benefits Working Group of the Council for Scientific Policy. Manchester: Manchester University, 1972. 169 leaves. Mimeo.

3.056 GIBBONS, M. and JOHNSTON, R.D.
'The role of science in technological innovation'. Research Policy, Vol.3, No.3, November 1974. pp.220-242.

3.057 GOLD, Bela
'Economic effects of technological innovations'. Management Science, 1964, Vol. 11(1), 105-134.

3.058 GOLDING, A.M.
The semi-conductor industry in Britain and the United States. A case study in innovation, growth and the diffusion of technology. D.Phil. Thesis, University of Sussex, 1971. 445 leaves.

3.059 GRANBERG, Anders and STANKIEWICZ, Rickard
The development of 'generic technologies' - the cognitive
aspects. Lund: Research   Policy   Institute, University of
Lund, 1981. 29p.

3.060 GREENBERG, Edward, HILL, Christopher T. and
       NEWBURGER, David J.
The influence of regulation and input costs on process
innovation: a case study of ammonia production. St. Louis:
Centre for   Development   Technology   Washington   University,
1977. 263p.

3.061 GRONHAUG, Kjell and FREDRIKSEN, Tor
Innovative conduct in the Norwegian textile industry: an
exploratory study. No imprint, 1978. 26p. (Notat nr. 13/78).

3.062 HAEEFNER, Erik
'The   innovation   process'. Technology Review,   March/April
1973, p. 18-25.

3.063 HAMILTON, David
'Products from inventions'. New Scientist, 9   January   1969,
Vol.41(631), 66-67.

3.064 HANLE, Paul A. (ed)
High technology on earth: studies in using aerospace
systems and methods. Washington DC: Smithsonian Institution
Press, 1979. 59p. (Smithsonian Studies in Air and Space, No.
3).

3.065 HARROD, Sir Roy Forbes
'The "neutrality" of improvements'. Economic Journal,   1961,
Vol.71(282), 300-304.

3.066 HARTLAND, John
Trident blind landing: a study in innovation.   A   thesis
submitted to the Victoria University of Manchester for the
degree of Master of Science, May, 1971. 110 leaves. Mimeo.

3.067 HAUSTEIN, Heinz-Deiter, MAIER, Harry and UHLMANN,
       Liutpold
Innovation and efficiency. Laxenburg, Austria: IIASA, 1981.
67p.

3.068 HAUSTEIN, Heinz-Dieter
Factor profiles of the innovation process as an analytical
tool for innovation policy. Presented at the IIASA Workshop
"Innovation policy and firm strategy", 4-6 December, 1979.
Laxenburg: I.I.A.S.A., 1979. 17p. (Working paper - 79-127).

3.069 HAVELOCK, Ronald G.
Planning for innovation: a comparative study of the
literature on the dissemination and utilization of
scientific knowledge. Ann Arbor: Centre   for   Research   on
Utilization of Scientific Knowledge, 1969.

3.070 HAWKES, Nigel
'Innovation and the universities'. Science Journal, March
1970, Vol.6(3), 75-79.

3.071 HAYES, Richard J.
A study of the transfer of technology from government
sponsored R & D to commercial operations in selected
electronic companies. Cambridge, Mass., Technology Utilization Office NASA Electronics Research Center, 1967.

3.072 HEAL, Geoffrey
Notes on the economic consequences of uncertain product
quality. Brighton: University of Sussex. 1973. 43 leaves.
(University of Sussex Economics seminar paper series 73/06).

3.073 HILDRED, William M. and BENGSTON, Leroy A.
Surveying investment in innovation. Denver: Denver Research
Institute, 1974. 144p.

3.074 HILL, Stephen C. and BELL, Martin R.
Paradigms and practice: innovation and technology transfer
models: their unexamined assumptions and inapplicability
outside developed countries. Preliminary   Draft.   Paper
presented at the Seminar on Technology Transfer at the Dept.
of Industrial Economics, University of Stirling, July 18-19,
1974. 84 leaves.

3.075 HOLLANDER, Samuel
The sources of increased efficiency: a study of Dupont
rayon plants. Cambridge (Mass.): M.I.T., 1965. 228p.

3.076 HOULTON, Robert
'The process of innovation: magnetic recording and the
broadcasting industry in the USA'. Bulletin of the Oxford
University Institute of Economics and Statistics,   February
1967, Vol. 29(1), 41-59.

3.077 ILLINOIS INSTITUTE OF TECHNOLOGY - Research
          Institute
Technology in retrospect and critical events in science.
Paper prepared for the National Science Foundation. 2 Vols.
(Chicago), 1968-69.

3.078 INDUSTRIAL RESEARCH INSTITUTE

Position statement on the impact of regulation on innovation. New York: I.R.I., 1979. 3p.

3.079   INDUSTRIAL RESEARCH INSTITUTE/RESEARCH CORPORATION
Contributions of basic research to recent successful industrial innovation. Prepared for the National Science Foundation. St. Louis: IRI/RC, 1979. Individually paginated. Mimeo. (PB80-160179)

3.080   INSTITUTE FOR INDUSTRIAL RESEARCH AND STANDARDS
The role of scientific and technical information in successful commercial innovation. By Liam Clifford. This report was commissioned by the OECD on behalf of the Information Policy Group. Dublin: I.I.R.S. 1975. 36 leaves. Mimeo.

3.081   INSTITUTE FOR INNOVATIVE ENTERPRISE
Incentives for innovation: a study of innovation in Maine industry - summary. Portland: University of Maine, 1976. 26p.

3.082   INSTITUTION OF MECHANICAL ENGINEERS.
Industrial innovation - the view and comments of the Institute of Mechanical Engineers on the report published by the Advisory Council for Applied Research and Development. London: The Institution, 1979. 10 leaves.

3.083   INTERNATIONAL LABOUR OFFICE
'A tabulation of case studies on technological change'. Labour and Automation Bulletin, No. 2, 1965. Whole issue.

3.084   INTERNATIONAL LABOUR ORGANISATION - Textile Committee
General report: recent events and developments in the textile industry. Ninth session of the Textiles Committee, Geneva, 1973. Report 1, pt.3. Geneva: I.L.O., 1973. 128p.

3.085   JACOBSSON, Staffan
'Electronics and the technology gap - the case of numerically controlled machine tools'. Bulletin of the I.D.S., Vol. 13, No. 2, March 1982. p. 42-46.

3.086   JENKINS, G.T. and DRINKWATER, H.G.
'Corfam versus cowhide: the complete case history'. The Director, May 1969, 277-283.

3.087   JERVIS, P. and SINCLAIR, T.C.
Conditions for successful technology transfer and innovation in the U.K. Paper given at the NATO Conference ASI-TT, June 24th - July 1973. 24 leaves.

3.088 JERVIS, Paul
'Innovation and technology transfer — the roles and characteristics of individuals. IEEE Transactions on Engineering Management, Vol. EM-22, No. 1, Feb. 1975, p. 19-27.

3.089 JEWKES, John and others
The Sources of Invention. 2nd ed. London, MacMillan, 1969.

3.090 JOHNSON, P.S.
The economics of invention and innovation: with a case study of the development of the hovercraft. London: Robertson, 1975. 329p.

3.091 JOHNSTON, Ron and GIBBONS, Michael
'Characteristics of information usage in technological innovation'. IEEE Transactions on Engineering Management, Vol. EM-22, No. 1, February 1975. p. 27-34.

3.092 JOHNSTONE, R. Edgeworth
'The nature of technological development'. Science Journal, November 1967, Vol. 3(11), 81-85.

3.093 JONES, Donald Bliss
A systems methodology for analysis of institutional involvement in the process of technological change. Submitted to the Graduate College of Texas, A & M University in partial fulfillment of the requirement for the degree of Doctor of Philosophy, May 1975. Ann Arbor, Michigan: University Microfilms International, 1980. 448p.

3.094 KAUFMAN, Morris
The history of P.V.C.: the chemistry and industrial production of polyvinyl chloride. London, Maclaren, 1969. 84p.

3.095 KELLY, Patrick
Technological innovation: a critical review of current knowledge. Atlanta, Georgia: Advanced Technology & Science Studies Group, Georgia Tec., 1975. Executive Summary 261 leaves. Vol. 1 – The ecology of innovation. 466p. Vol. 2 – Aspects of technological innovation: commissioned papers. 541p. Vol. 3 – Cross-classifications of bibliographical data base. 711p. Vol. 4 – Selected literature abstracts 452p.

3.096 KENNEDY, Charles
'Induced bias in innovation and the theory of distribution'. Economic Journal, 1964, Vol. 74(295), 541-547.

3.097 KLEIMAN, Herbert S.

The integrated circuit: a case study of product innovation in the electronics industry. Thesis, George Washington University, D.E.A., 1966. (Ann Arbor, University Microfilms, 1969).

3.098 KOLODNEY, Robert and PEPINO, Gabriel
'Obstacles to innovation'. European Business, October 1968, No. 19, 14-17.

3.099 LANGRISH, J.
Case studies of ten innovations which earned the Queen's award to industry, 1966. Typescript. Paper read to the British Association for the Advancement of Science, Leeds, 1967.

3.100 LANGRISH, J. et al.
Wealth from knowledge: studies of innovation in industry. London: MacMillan, 1972. 477p.

3.101 LAPP R.E.
An empirical study of some relationships between techno-logical innovations and organizational characteristics in eight railroads. A thesis submitted to the graduate school in fulfilment of the degree of Master of Science in the field of transportation. Evanston: Northwestern University, 1966. 169p.

3.102 LAWRENCE, P.J.
'The motor with the hole - invention and innovation'. Advancement of Science, December 1969, 120-152.

3.103 LAWSON, William D. and others
'Corfam: research brings chemistry to footwear'. Research Management, January 1965, Vol. 8(1), 5-26.

3.104 LAYTON, Christopher, HARLOW, Christopher and DE
        HOUGHTON, Charles
Ten innovations: an international study on technological development and the use of qualified scientists and engineers in ten industries. London: Allen & Unwin, 1972. 199p.

3.105 LEPKOWSKI, Wil
'Long-term economic slide inhibiting innovation'. Chemical and Engineering News, Vol. 59, No. 43, 26th October 1981, p. 20-21.

3.106 LEVIN, Richard C.
Innovation in the semiconductor industry: is a slowdown imminent? New Haven, Conn.: Yale University, 1981. 24

leaves. Mimeo.

3.107 LIEBERMAN, Marvin B.
The links from science to technology - a look at
electronics. May 12, 1975. 54 leaves. Mimeo.

3.108 LORENZ, Christopher
Investing in success: how to profit from design and
innovation. London: The Anglo-German Foundation for the
Study of Industrial Society, 1979. 26p.

3.109 LUTTEMAN, Helena Dahlback
'Industrial design in Sweden'. Current Sweden, No. 250 (6),
May 1980. Whole issue.

3.110 LYNN, Leonard H.
How Japan innovates: a comparison with the U.S. in the case
of oxygen steelmaking. Boulder, Colorado: Westview Press,
1982. 211p. (A Westview Replica Edition).

3.111 MADDOCK, I.
'Inventiveness and innovation in industry: nothing left but
our brains'. An account of a symposium held on 5th Sept.,
1969, Exeter. Advancement of Science, December 1969, p.
107-119.

3.112 MAHER, Arthur
A study of innovation in the United Kingdom. Plymouth: The
Plymouth Business School, Plymouth Polytechnic, 1982. 106p.
Mimeo.

3.113 MAIER, Frank H.
An economic analysis of adoption of the mechanical cotton
picker. Ph.D. thesis, University of Chicago, 1969. PB 184
320.

3.114 MAIER, Harry and HAUSTEIN, Heinz-Dieter
'Innovation, efficiency cycle and strategy implications',
Technological Forecasting and Social Change, Vol.17, 1980,
pp.35-49.

3.115 MAJOR, Randolph T.
'Cooperation of science and industry in the development of
the antibiotics'. Chemical and Engineering News, 25 October
1948, Vol. 26(43) 3187-3190.

3.116 MANSFIELD, Edwin et al
Social and private rates of return from industrial
innovations. No imprint, 1974/5. 28 leaves. Mimeo.

3.117 MARQUIS, Donald G.
'The anatomy of sucessful innovations', Innovation, No. 7, 1969, p. 28-35.

3.118 MARTIN, Ben R. and IRVINE, John
'Spin-otf from basic science: the case of radioastronomy', Physics in Technology, Vol. 13, September 1981. p. 204-212.

3.119 MERTON, Robert K.
'Fluctuations in the rate of industrial invention', Quarterly Journal of Economics, 1935, Vol. 49(3), 454-474.

3.120 METCALFE, J.S.
'The rewards and problems of successful technological innovation', Metals & Materials, Feb 1970, 63-67.

3.121 MICHAELIS, Michael
'The environment for innovation', European Business, January 1969, No. 20, 10-16.

3.122 MILES, Caroline
Lancashire textiles: a case study of industrial change.
Cambridge: C.U.P., 1968. 124p. (N.I.E.S.R. Occasional Papers, XXIII)

3.123 MILLER, Albert W. and HAINES, George E.
'A market study of a new industrial product', IMRA Journal, Vol. 6, No. 4, P. 195-205, Nov. 1970.

3.124 MUELLER, Willard F.
'A case study of product discovery and innovation costs', Southern Economic Journal, July 1957, Vol. 24(1), 80-86.

3.125 MULLER, Peter
Austria innovative. Muenchen: Jugend and Volk, 1979. 103p.

3.126 MaCDONALD, Stuart and BRAUN, Ernest
'The transistor and attitude to change.' American Journal of Physics, Vol. 45, No. 11, November 1977, p. 1061-1065.

3.127 MacLAURIN, W. Rupert
Invention and innovation in the radio industry. New York, MacMillan, 1949.

3.128 MacLAURIN, W. Rupert
'The sequence from invention to innovation and its relation to economic growth'. Quarterly Journal of Economics, February 1953, Vol. 67(1), 97-111.

3.129 McADAMS, Alan K.

'Big steel, invention and innovation, reconsidered'. Quarterly Journal of Economics, August 1967, Vol. 81(3), 456-482.

3.130 McCARTHY, M.C.
Patterns of innovation in the chemical industry. A thesis presented for the degree of Ph.d. in the University of Stirling, 1971. Dept of Industrial Science, Stirling University. 220 leaves.

3.131 NATIONAL ACADEMY OF ENGINEERING.
The process of technological innovation Symposium sponsored by the National Academy of Engineering, April 24, 1968. Washington: National Academy of Sciences, 1969. NAS Publication 1726.

3.132 NATIONAL SCIENCE FOUNDATION
The influence of regulation and input costs on process innovation. Interim report: innovation in ammonia production, by E. Greenberg (and others), for the National Science Foundation. St. Louis: Washington University, 1976. 183p.

3.133 NELSON, Richard R.
The role of industrial innovation in the United States economy. To be presented at a forum organized by the Congressional Research Service, Washington D.C., 18 June 1980. New Haven, Connecticut: Yale University, 1980. 12 leaves. Mimeo.

3.134 NELSON, Richard R. and WINTER, Sidney G.
In search of a theory of innovation. Paper presented at the BMFT/NSF Seminar on Technological Innovation, Bonn, 5-9 April 1976. 53 leaves. Mimeo.

3.135 NORDHAUS, William D.
'Some sceptical thoughts on the theory of induced innovation'. Quarterly Journal of Economics, Vol.87, May, 1973, pp.208-219.

3.136 NYSTROM, Harry
Creativity and innovation. Chichester: Wiley, 1979. 125p.

3.137 PARKINSON, D.
Monoclonal antibodies: another success for biotechnology? A case study of a rapidly growing field of biotechnology, its history, development and commercial opportunities. Prepared for the FAST Programme, sub-programme Biosociety. Brussels: Commission for the EEC, 1981. 165p. Mimeo. (Cl.4 Analysis of the Scientific Disciplines germane to Biotechnology).

3.138 PARSONS, S.A.J.
The framework of technical innovation. London, Macmillan,
1968.

3.139 PAVITT, Keith
'Research, innovation and economic growth'. Nature, 1963,
200(4903), 206-210.

3.140 PAVITT, Keith and ROTHWELL, Roy
'Feedback: a comment on "A dynamic model of process and
product innovation"'. Omega, Vol.4, No.4, 1976, p.375-377.

3.141 PIERCE, John R.
'Innovation in technology'. Scientific American, 199(3),
1958, pp.117-130.

3.142 PRAGIER, Gerald
Pharmaceutical innovation: three case studies of corporate
achievement meriting the Queen's Award to Industry.
Dissertation submitted in part fulfilment of the
requirements for the Degree of Master of Science at the
Victoria University of Manchester, 1974. 186 leaves.
Photocopy.

3.143 PRICE, Derek de Solla
The relations between science and technology and their
implications for policy formation. Lecture given at the
Royal Institute of Technology, Stockholm, 6 June 1972. 32p.

3.144 PRICE, William J. and BASS, Lawrence W.
'Scientific research and the innovative process'. Science,
16 May 1969, 164(3881), 802-806.

3.145 RAY, George F.
The innovation process in the energy industries. Cambridge:
C.U.P. 120p. (NIESR occasional papers XXX).

3.146 RAY, George F.
'Innovation in industry: the state and the results of
recent economic research in Western European countries
except F.R.Germany'. Research Policy, Vol.3, No.4, January
1975. pp.338-359.

3.147 RAYMENT, Judith J.
The rise and fall of artificial leather: critical
literature survey. Spring 1975. 25 leaves. Mimeo.

3.148 ROBERTS, Edward B.
'A basic study of innovators; how to keep and capitalize on
their talents'. Research Management, 1968, 11(4), 249-266.

3.149 ROBERTS, Robert E. and ROMINE, Charles A.
Investment in innovation: a study of innovation and the
feasibility of measuring resources devoted to it - final
report. Kansas City: Midwest Research Institute, 1974. 131
leaves. Mimeo.

3.150 ROBERTSON, Andrew
'The innovation process: success and failure'. The
Business Economist, 1972, 4(2), pp.71-81.

3.151 ROBERTSON, Andrew
The lessons of failure: cases and comments on consumer
product innovation. London: Macdonald, 1974. 106p.

3.152 ROBERTSON, Andrew
'Multinational expertise is not enough: the sobering story
of Corfam'. Multinational Business, No.2, 1975, pp.1-7.

3.153 ROBERTSON, Andrew and FROST, Michael
'Duopoly in the scientific instrument industry: the milk
analyser case'. Research Policy, Vol.7, No.3, July 1978,
pp.292-316.

3.154 ROBERTSON, Thomas S.
Innovative behaviour and communication. New York, London:
Holt, Rinehart and Winston, 1971. 331p.

3.155 ROTHWELL, Roy
'Technological innovation in textile machinery: the role of
radical and incremental technical change'. Textile
Institute and Industry. November 1976, pp.330-336.

3.156 ROTHWELL, Roy
'Innovation in machinery: some factors in success and
failure'. Textile Manufacturer & Knitting World, 1976, No.3,
pp.47-49 and 65.

3.157 ROTHWELL, Roy
'Picanol Weefautomaten: a case study of a successful
textile machinery builder'. Textile Institute and Industry,
March 1976. pp.103-106.

3.158 ROTHWELL, Roy and ZEGVELD, Walter
'Possibilities for innovation in small and medium sized
manufacturing firms (s.m.f's.)'. IEE Proceedings Vol.127,
Part A, No.4, May 1980, pp.267-271.

3.159 ROY, Robin, WALKER, David and WALSH, Vivien
Product design, innovation and competitiveness in British
manufacturing industry: aims and objectives of an

investigation. Milton Keynes: The Open University, Design Innovation Group, May 1980. 3 leaves. Mimeo.

3.160 RUBENSTEIN, Albert H., CHAKRABARTI, Alok K. and O'KEEFE, D.
Final technical report on field studies of the techno-logical innovation process.   Evanston: Northwestern University, 1974. 157p.

3.161 RUBY, Bjarne
Design for innovation: a cybernetic approach.   Copenhagen: Institute for Future Studies, 1974. 111p.

3.162 SABATO, Jorge A.
'Using science to "manufacture" technology'. Impact of Science on Society, Vol.XXV, No.1, January-March 1975, pp.37-44.

3.163 SAHAL, Devendra
'Models of technological development and their relevance to advances in transportation'. Technological Forecasting and Social Change, Vol.16, 1980, pp.209-227.

3.164 SAHAL, Devendra
'Alternative conceptions of technology'. Research Policy Vol.10, 1981, pp.2-24.

3.165 SAHAL, Devendra
'The farm tractor and the nature of technological innovation'. Research Policy Vol.10, No.4, October 1981, pp.368-402.

3.166 SAHAL, Devendra
Patterns of technological innovation.   Reading,   Mass.: Addison-Wesley, 1981. 381p.

3.167 SAMUELSON, Paul A.
'A theory of induced innovation along Kennedy-Neisacker lines'. The Review of Economics and Statistics 47, (4), November 1965. pp.343-356.

3.168 SCHERER, F.M.
'Invention and innovation in the Watt-Boulton steam engine venture'. Technology and Culture, Vol.6, No.2, Spring 1965, pp.165-187.

3.169 SCHERER, F.M.
'Demand-pull and technological invention: Schmookler revisited'. The Journal of Industrial Economics, Vol.XXX, No.3, March 1982, pp.225-237.

3.170 SCHMOOKLER, Jacob
'Economic sources of inventive activity'. Journal of Economic History 22(1), March 1962, pp.1-20.

3.171 SCHMOOKLER, Jacob
Invention and economic growth. Cambridge, Mass., Harvard U.P, 1966.

3.172 SCHMOOKLER, Jacob and BROWNLEE, Oswald
'Determinants of inventive activity'. American Economic Review 52(2), 1962, pp. 165-176.

3.173 SCHOTT, Kerry
Industrial innovation in the United Kingdom, Canada and the United States. London: British-North American Committee, 1981. 65p.

3.174 SCHWARTZMAN, David
Innovation in the pharmaceutical industry. Baltimore, London: The John Hopkins University Press, 1976. 399p. Pbk.

3.175 SEETZEN, J. and FERRARI, A.
Results of innovation research in USA and Europe. International Institute for the Management of Technology. 18 leaves.

3.176 SHELL, Karl
'Toward a theory of inventive activity and capital accumulation'. American Economic Review, 56(2), 1966, pp.62-69.

3.177 SHELTON, John P.
'Special problems of innovation and technology'. Idea, Vol.15, No.2, 1971, pp.281-292.

3.178 SIEGEL, Irving H.
'The role of scientific research in stimulating economic progress'. American Economic Review, 50(2), 1960, pp.340-345.

3.179 SMITH, Thomas M. and MASON, Dwayne R.
Final report on a case study on the relation of basic research to advances in technology. Mimeo. Norman, Oklahoma, University of Oklahoma Research Institute, 1968.

3.180 STANFORD RESEARCH INSTITUTE
Case studies and modelling of the innovation process. Research proposal submitted to the National Science Foundation. Menlo Park, California: SRI, 1973. 31 leaves. Mimeo.

3.181 STONEMAN, P.
Product and process innovation and the R & D market
structure relationship. University of Warwick, 1979. 18
leaves. Mimeo.

3.182 STUART, George F.
Technology and innovation in the New Zealand manufacturing
industry - empirical studies. Resource Paper 1. Lower Hutt,
New Zealand: Physics and Engineering Laboratory, DSIR,
1980. 152p. Mimeo. (PEL Report No.685).

3.183 STUART, George F. and McCULLOCH, D.G.
Technology and innovation in the New Zealand manufacturing
industry - a discusson paper. Lower Hutt, New Zealand:
Physics and Engineering Laboratory, OSIR, 1980. 31p. Mimeo.
(OSIR Discussion paper No.4).

3.184 STUBBS, Peter
Innovation and research: a study in Australian industry.
Melbourne: F.W. Cheshire for the Institute of Applied
Economic Research, University of Melbourne, 1968. 232p.

3.185 TEUBAL, Morris
On user needs and need determination: aspects of the theory
of technological innovation. Jerusalem: Maurice Falk
Institute for Economic Research in Israel, 1977. 38p.
(Discussion paper no.774).

3.186 TEUBAL, Morris
Need determination, product type and the inducements to
innovate. Jerusalem: Maurice Falk Institute for Economic
Research, 1979. 26p. Discussion paper no.791).

3.187 TOWNSEND, J.F.
Innovation in coal mining machinery: "The Anderton Shearer
Loader" - the role of the NCB and the supply industry in its
development. Brighton: SPRU, 1976. 84p. (SPRU Occasional
Paper series No.3).

3.188 UNITED STATES - National Science Foundation
Successful industrial innovations: a study of factors
underlying innovation in selected firms. By Sumner Myers and
Donald G. Marquis. Washington: USGPO, 1969.

3.189 UNITED STATES - Office of the Director of Defense
          Research and Engineering
First interim report on Project Hindsight (summary). By C.W.
Sherwin and R.S. Isenson. Washington 1966.

3.190 UTTERBACK, James M.

What are the systems for innovation: micro and macro. Paper presented at the American Chemical Society and Industrial Research Institute Symposium on Innovation in Industry, Washington D.C.: September 3-5, 1979. Washington D.C.: ACS/IRI, 1979. No pagination. Mimeo.

3.191 UTTERBACK, James M., ALLEN, Thomas J., HOLLOMON, J.
       Herbert and SIRBU, Marvin A.
'The process of innovation in five industries in Europe and Japan'. IEEE Transactions in Engineering Management, Vol. EM-23, No.1, February 1976, pp.1-9.

3.192 VON HIPPEL, Eric
The dominant role of users in the scientific instrument innovation process. Cambridge, Mass.: Alfred P. Sloan School of Management, 1975. 43 leaves. Mimeo.

3.193 VON HIPPEL, Eric
Industrial innovation by users: evidence, explanatory hypotheses and implications. Cambridge, Mass.: Alfred P. Sloan School of Management, 1977. 29 leaves. (Working pper 953-77).

3.194 VON HIPPEL, Eric and FINKELSTEIN, Stan N.
Product designs which encourage - or discourage - related innovation by users: an analysis of innovation in automated clinical chemistry analyzers. Cambridge, Mass.: MIT, 1978. 23p. Alfred P. Sloan School of Management. (WP 1011-78).

3.195 WIERZYNSKI, Gregory H.
'The eccentric lords of float glass'. Fortune, 78(1), July 1968, 90-92, pp.121-124.

3.196 WYATT, G.J.
The supply of inventions. presented at a workshop on the Causes and Consequences of Technological Change at Loughborough University of Technology, April 1981. Edinburgh: Heriot-Watt University, Department of Economics, 1981. 20p and Appendices. Mimeo.

3.197 YIN, Robert K. et al
A review of case studies of technological innovations in state and local services. Santa Monica: The Rand Corporation, 1976. 310p.

3.198 de BRESSON, Christian and TOWNSEND, J.
'Multivariate models for innovation - looking at the Abernathy-Utterback model with other data'. Omega, Vol.9, No.4, 1981, p. 429-436.

3.199   de MELTO, Dennis P. and McMULLEN, Kathryn E. and
        WILLS, Russel M>
Innovation and technological change in five Canadian
industries. Preliminary report. Ottawa: Economic Council of
Canada, 1980. 309p. Mimeo. (Discussion Paper No. 176)

3.b Diffusion Of Innovations: Firms & Countries

3.200 ACHILLADELIS, B., ROBERTSON, A.B. and FREEMAN, C.
'A study on   innovation in   the   chemical   industry:
preliminary   report   on   project   "Sappho"'.
Chemistry and Industry, No.10, 6 March 1971, pp.269-273.

3.201 ACHILLADELIS, Basil G.
Process innovation in the chemical industry.   A   thesis
presented for the degree of Doctor of Philosophy in History
and Social Studies of Science at the University of Sussex.
Sept. 1973, 231p.

3.202 ARCHER, John F.
The diffusion of space technology by means of technical
publications: a report based on the distribution, use and
effectiveness of "selected welding techniques". Washington,
NASA, Scientific and Technical Information Facility, 1964.
X66-35850.

3.203 BAR-ZAKAY, Samuel N.
'Technology transfer model'. Technological Forecasting and
Social Change, Vol. 2, 1971, p.321-337.

3.204 BAUMANN, H.G.
The diffusion of the basic oxygen process in U.S. and
Canadian steel industries, 1955-1969. University of Western
Ontario, 1973. 26 Leaves. (Research report no. 7303).

3.205 BEALS, Ralph L.
Problems of resistance and adaptation to technological
change. Unesco, E/CONF39/J/68, 1962. United Nations
Conference on: "The application of science and technology
for the benefit of the less developed areas".

3.206 BRADBURY, Frank, JERVIS, P., JOHNSTON, R. and
        PEARSON, A. (eds)
Transfer processes in technical change. Alphen aan den Rijn,
The Netherlands: Sijthoff & Noordhoff, 1978. 280p.

3.207 BROWN, Lawrence

Models for spatial diffusion research - a review.   Mimeo.
Evanston,   III.,   Department   of   Geography,   Northwestern
University, 1965. Research Report No. 10.

3.208 BROWN, Lawrence
Diffusion dynamics: a review and revision of the
quantitative theory of the spatial diffusion of innovation.
Lund, Gleerup, 1968. Lund Studies in Geography, Series D,
No. 29.

3.209 BROWN, Lawrence
Innovation diffusion: a new perspective. London: Methuen,
1981. 345p.

3.210 CARTER, Charles F. and WILLIAMS, Bruce R.
Industry and technical progress: factors governing the
speed of application of science. London, OUP, 1957.

3.211 CENTRE FOR THE STUDY OF INDUSTRIAL INNOVATION
Report for the National Economic Development Office on the
level of innovation activity in the pharmaceutical industry.
London: The Centre for the Study of Industrial Innovation,
1971. 48p.

3.212 CHAKRABARTI, Alok Kumar
The effects of techno-economic and organizational factors
on the adoption of NASA-innovations by commercial firms in
the U.S. A dissertation submitted in partial fulfilment of
the requirements for the degree of Doctor of Philosophy.
Evanston, Illinois: Northwestern University, 1972.   398p.
(Program of Research on the Management of Research &
Development - Project 8 (1)).

3.213 COLE, Sam
Income distribution and trade in a model of innovation and
transfer of technology. Brighton: SPRU, 1979. 16p. Mimeo.

3.214 COLEMAN, J.S.
Medical innovation: a diffusion study. Indianapolis, Bobbs-
Merrill, 1966.

3.215 COOPER, Michael H.
Patents, innovation and exports in Britain, India, Italy,
Japan and U.S.A. Office of Health Economics, Winter lecture
1967. Typescript, 1967.

3.216 CURNOW, Ray C.
The computer innovation cycle. Paper read at the European
Conference on Technological Forecasting, University of
Strathclyde, 1968.

3.217 CURNOW, Ray C.
The innovation cycle in the manufacture and application of computer systems. In: European Conference on Technological Forecasting, Glasgow, June 24th-26th, 1968. Typescripts, 1968.

3.218 DANHOF, Clarence H.
Technology transfer by people transfer: a case study. Washington: George Washington University, 1969. Program of Policy Studies in Science and Technology, Staff Discussion Paper 403.

3.219 DAVIES, Stephen
The diffusion of process innovations. Cambridge: CUP, 1979. 193.

3.220 DENNY, M. et al.
Estimating the effects of diffusion of technological innovations in telecommunications: the production structure of Bell Canada. University of Toronto, Institute for Policy Analysis, 1979. 33p (Working paper no.7913).

3.221 FELLER, Irwin and MENZEL, Donald C.
Diffusion milieux as a focus of research on innovation in the public sector. Paper prepared for the 1975 Annual meeting of the American Political Science Association, San Francisco, September 2-5, 1975. No imprint. 43 leaves. Mimeo.

3.222 FELLER, Irwin, MENZEL, Donald and ENGEL, Alfred J.
Diffusion of technology in state mission-oriented agencies. Pennsylvania State University, Center for the Study of Science Policy, 1974. 300p.

3.223 FLIEGEL, Frederick C. and KIVLIN, Joseph E.
'Attributes of innovations as factors in diffusion', American Journal of Sociology, November 1966, Vol.72(3), p.p.235-248.

3.224 GILBERT, E.W.N.
The investigation of the diffusion of numerical control system innovations into the engineering industries of the United Kingdom and Scandinavia. Typescript. 1969.

3.225 GLOBERMAN, Steven
'A study of new technology adoption by Canadian textile firms'. The Clemson University Review of Industrial Management and Textile Science, Vol. 14, (1), Spring 1975. pp.33-42.

3.226 GLOBERMAN, Steven
'Technology diffusion in the Canadian carpet industry'.
Research Policy, Vol. 4, 1975. pp. 192-206.

3.227 GOLD, Bela
'Technological diffusion in industry: research needs and
shortcomings'. Journal of Industrial Economics, Vol.XXIX,
No.3, March 1981. pp.247-269.

3.228 GREASLEY, David G.
The diffusion of a technology: the case of machine coal
cutting in Great Britain, 1900-1938. Thesis presented for
the Degree of Ph.d. of the University of Liverpool, May
1979. Liverpool: University of Liverpool, 1979. 296 leaves.

3.229 HAGERSTRAND, Torsten
Innovation diffusion as a spatial process. London, University of Chicago Press, 1967. First published 1953.

3.230 HAUSTEIN, Heinz-Dieter and MAIER, Harry
The diffusion of flexible automation and robots. Laxenburg:
I.I.A.S.A., 1981. 44p. (Working paper - 81-152).

3.231 HEDEN, Carl-Goran
The potential impact of microbiology on developing
countries. Vienna: UNIDO, 1981. 66p. Mimeo. (UNIDO/IS.261).

3.232 HISLOP, Drummond and HOWES, Michael
The transfer of technology to the Thai silk industry.
Applied Scientific Research Corporation of Thailand,
September 1974. Mimeo.

3.233 HOUGH, Granville W.
Technology diffusion: federal programs and procedures. Nt.
Airy: Lomond Books, 1975. 406p.

3.234 IWAUCHI, Ryoichi
'Adaption to technological change'. The Developing
Economies, Vol.7, No. 4, December 1969, p. 428-450.

3.235 JEREMY, David J.
Transatlantic industrial revolution: the diffusion of
textile technologies between Britain and America,
1790-1830's. Oxford: Blackwell, 1981. 384p.

3.236 KATZ, Elihu
The diffusion of innovation: an analysis with case studies.
New York, Wiley, 1962.

3.237 KIMBERLY, John R.

Hospital adoption of innovation in medical and managerial technology: individual, organizational and contextual/environment effects - final project report. Yale University, 1978. 85 leaves.

3.238 KRAUSHAR, Peter M.
New products and diversification. 2nd ed. London: Business Books, 1977. 212p.

3.239 LEONARD-BARTON, Dorothy and ROGERS, Everett M.
Horizontal diffusion of innovations: an alternative paradigm to the classical diffusion model. Cambridge, Mass.: MIT, 1981. 23 leaves. Mimeo. (Alfred P. Sloan School of Management Working Paper No. 1214-81)

3.240 MACKERRON, G. and THOMAS, S.D.
Industry and regulatory responses to problems in nuclear power development. Final report to Science and Engineering Research Council, Brighton: SPRU, 1981.

3.241 MADDALA, G.S. and KNIGHT, Peter T.
'International diffusion of technical change - a case study of the oxygen steel making process'. Economic Journal, September 1967, Vol. 77(307), 531-558.

3.242 MANSFIELD, Edwin
'Technical change and the rate of imitation'. Econometrica, 1961, Vol. 29(4), 741-766.

3.243 MANSFIELD, Edwin
'Intrafirm rates of diffusion of an innovation.' Review of Economics and Statistics, November 1963, Vol. 45(4), 348-359.

3.244 MANSFIELD, Edwin
Determinants of the speed of application of new technology. Mimeo. Paper given at meeting of the International Economic Association, St. Anton, Austria, 1971.

3.245 METCALFE, J.S.
The diffusion of innovation and technical progressiveness in the Lancashire textile industry. M.Sc. thesis, Victoria University of Manchester, 1968.

3.246 MULLINS, L.
'Developments in the industrial use of natural rubber', Polymer Journal, (Singapore), Vol. 1, p. 27-29, 1970.

3.247 MYERS, Sumner
'Technology transfer and industrial innovation', Looking

Ahead, Febuary 1967, Vol. 15(1), 1-7.

3.248 NABSETH, L. and EAY, G. F. (eds)
The diffusion of new industrial processes: an international
study. London: C.U.P. 1974. 324p.

3.249 NATIONAL ACADEMY OF ENGINEERING.
Technology transfer and utilization recommendations for
redirecting the emphasis and correting imbalance.
Washington, D.C.: National Academy of Engineering,1974.
Microfiche. (PB 232 123)

3.250 NICOSIA, F.M.
Technology change, product proliferation, and consumer
decision processes. Volume 1 - summary report, technology
and consumers: individual and social choices. August 1974.
88p. A cooperative research project of the University of
California, Berkeley and Columbia University, New York.

3.251 NIEHOFF, Arthur H. and ANDERSON, J. Charnel
'The process of cross-cultural innovation'. International
Development Review, June 1964, 6, 5-11.

3.252 PAVITT, Keith and SOETE Luc
Comment on G. Lorenz's paper - "The diffusion of emerging
technology among industrial countries: direction, channels &
speed". Paper presented at the Conference on Emerging
Technology: consequences for economic growth, structural
change and employment in advanced open economies. Kiel, June
24-26, 1981. (No imprint). 7 leaves.

3.253 PECK, Stephen C. and STONER, Robert D.
The diffusion of technological innovations among privately
owned electric utilities, 1950-1975. A report to the NSF. No
imprint. 1977. 149 leaves. Mimeo.

3.254 PIRA
Developments in printing technology and other technologies
which may affect the printer: a ten year forecast.
Leatherhead: Pira, 1979. Individually paginated. (Prepared
by Pira for the Printing and Publishing Industry Training
Board).

3.255 POLYTECHNIC OF CENTRAL LONDON
The diffusion of technological innovation: a study of the
adoption of the electronic digital computer in process
control. October, 1973.

3.256 RADNOR, Michael, FELLER, Irwin and ROGERS, Everett
         (eds)

The diffusion of innovations: an assessment. Evanston, Illinois: Center for the Interdisciplinary Study of Science and Technology, Northwestern University, 1978. Individually paginated. Mimeo.

3.257 RAQUET, John W.E.
The impact of genetic manipulation on the chemical industry. Guildford: University of Surrey, 1981. 74 leaves. Mimeo.

3.258 RAY, George F.
'The diffusion of a new technology: a study of ten processes in nine industries'. National Institute Economic Review, May 1969, No.48, 40-83.

3.259 RAY, George F.
'A case study in the diffusion of a new technological process: the application of gibberellic acid in the malting process'. Birmingham University Chemical Engineer, 1970, 21(2), 40-44.

3.260 RICHARDSON, Peter R.
The acquisition of new process technology by firms in the Canadian mineral industries. Submitted in partial fulfillment of the requirements for the degree of Doctor of Philosophy, Faculty of Graduate Studies, The University of Western Ontario, London, Canada, April 1975. London, Ontario: University of Western Ontario, 1975. 199 leaves. Mimeo.

3.261 ROBERGE, Roger A., RAY, D.M. and VILLENEUVE, Paul Y.
'Invention, diffusion and allometry: a study of the growth and form of the pulp and paper industry in Central Canada'. Chapter 5 in: HAMILTON, F.E. Ian (ed) Spatial perspective on industrial organisation and decision making. London: John Wiley & Sons, 1975. 143-168p.

3.262 ROBERTS, Edward B. et al (eds)
Biomedical innovation. Cambridge, Mass./London, MIT Press, 1981. 395p. (MIT Series in Health and Public Policy. General Editor: Jeffrey E. Harris).

3.263 ROBSON, M. and CHESSHIRE, J.
Industrial energy demand: economic and technical change in the U.K. boiler stock. SPRU Occasional Paper Series No. 19. Brighton: SPRU, 1983.

3.264 ROGERS, Everett M.
Diffusion of innovations. New York, Free Press, 1962.

3.265 ROGERS, Everett M. with FLOYD, F.

Communication of innovations: a cross-culture approach.
Shoemaker. Second Edition. London: Collier-Macmillan, 1971.
476p.

3.266 ROMEO, Anthony A.
'Interindustry and interfirm differences in the rate of
diffusion of an innovation'. The Review of Economics and
Statistics,Vol.LVII, No.3, August 1975, pp.311-319.

3.267 ROMEO, Anthony A.
'The rate of imitation of a capital-embodied process
innovation'. Economica,Vol.44, February 1977, pp.63-69.

3.268 ROSENBERG, Nathan
'On technological expectations'. The Economic Journal, 86,
September, 1976, pp.532-535. (Center for Research in
Economic Growth Reprint No.145).

3.269 ROSENBERG, Nathan
Learning by using. Stanford University, 1979. 32 leaves.
Mimeo.

3.270 RUSH, H.J. MACKERRON,G and SURREY A.J.
'The advanced gas-cooled reactor: a case study in reactor
choice', Energy Policy, Vol.5, No.2, June 1977.

3.271 RUSH, H.J. and HOFFMAN, H.K.
Microelectronics and the garment industry: not yet a
perfect fit. Paper presented to an international conference
on Informatics and Industrial Development, 9-13 March 1981,
Trinity College, Dublin. Brighton: SPRU, April 1981. 13
leaves. Mimeo.

3.272 RUSH, Howard
'The rag trade threads its way into the micro age'.
Electronics Times No.99, 16 October 1980, pp.24-25.

3.273 RUSSELL, Louise B. and BURKE, Carol S.
Technological diffusion in the hospital sector. Washington
D.C.: National Planning Association, 1975. 194 leaves.

3.274 SAHAL, Devendra
'The temporal and spatial aspects of diffusion of
technology'. IEEE Transactions on Systems, Man and Cyber-
netics, Vol.SMC-9, No.12, December 1979, pp.829-839.

3.275 SAHAL, Devendra
'A theory of evolution of technology'. The International
Journal of Systems Science, Vol.10, No.3, 1979, pp.259-274.

3.276 SCOTT, Thomas William Keillor
Diffusion of new technology in the British and West German
carpet manufacturing industries: the case of the tufting
process. D.Phil. thesis presented at the University of
Sussex. 308 leaves. Mimeo.

3.277 SENKER, P. SCIBERRAS, E. SWORDS, N. and HUGGETT, C.
Forklift trucks: a study of a sector of the U.K.
Engineering Industry. A report prepared for the Engineering
Industry Training Board. Brighton: SPRU, September 1977.
192p. Mimeo.

3.278 SHEN, T.Y.
'Innovation, diffusion, and productivity changes'. Review
of Economics and Statistics, 43, 1961, pp.175-181.

3.279 SHIMWELL, Charles and WHITE-HUNT, Keith
The application and utilisation of computer and
microelectronic technology within manufacturing and business
systems with particular reference to the steel industry.
Stirling: Technological Economics Research Unit, Department
of Management Science and Technology Studies, University of
Stirling, 1979. 92p. Mimeo. (Discussion Paper No.22).

3.280 SHISHKO, Robert
Technological change through product improvement in
aircraft turbine engines. Santa Monica, California: Rand,
1973. 78p.

3.281 SIEGEL, Irving H. and WEINBERG, Edgar
'Numerical-control technology: antecedents, development,
diffusion'. Idea, 10(2), 1966, pp.129-142.

3.282 SMITH, R.J.
'The weaving of cotton and allied textiles in Great Britain:
an industry survey with special reference to the diffusion
of shuttleless looms'. National Institute Economic Review,
No.53, August 1970, pp.54-69.

3.283 STANFIELD, J.D. and others
Simulation of innovation diffusion. East Lansing, Michigan,
Department of Communication, Michigan State University,
1965. Working Paper 7. Mimeo.

3.284 STONEMAN, P.
On the change in technique: a study of the spread of
computer usage in the U.K., 1954-1970.   A   dissertation
submitted for the degree of Doctor of Philosophy in the
University of Cambridge. University of Cambridge, 1973.
317p. Mimeo.

3.285 STONEMAN, P.
Technological diffusion and the computer revolution: the
U.K. experience. Cambridge: C.U.P., 1976. 219p. (University
of Cambridge, Department of Applied Economics, Monograph
25).

3.286 STONEMAN, P.
'Embodiment of technological change and the speed of
adjustment'. Economica, Vol. 44, November 1977, pp.421-22.

3.287 STONEMAN, P.
Intra-firm diffusion, Bayesian learning and profitability.
University of Warwick, 1979. 21 leaves. Mimeo.

3.288 STONEMAN, P.
The rate of imitation, learning and profitability.
University of Warwick, 1979. 22 leaves. Mimeo.

3.289 STUART, George F.
Technology and its licencing. Paper presented to the
Chemical Engineering Group Symposium on "Technology: Buying
or Generating" at NZIE Conference, Hamilton, 14 February
1978. Lower Hutt, New Zealand: Physics and Engineering
Laboratory, OSIR, 1978. 9p. Mimeo.

3.290 SURREY, A.J.
'The future growth of nuclear power'. Energy Policy, Vol.1,
No.2, September 1973 and Vol.1, No.3, December 1973.

3.291 SURREY, A.J. BUCKLEY, C.M. and ROBSON, M.
'Heavy electrical plant'. in K.L.R. Pavitt(ed.), Technical
innovation and British economic performance, London: Macmillan, 1980.

3.292 SUTHERLAND, Alister
'Diffusion of an innovation in cotton spinning'. Journal
of Industrial Economics, 7(2), March 1959, pp.118-135.

3.293 THE CONTROL AND AUTOMATION MANUFACTURERS ASSOCIATION
A guide to the procurement of complex electronic control
and supervisory systems. London: The Control and Automation
Manufacturers Association. 1974. 35p. Mimeo.

3.294 TILTON, John E.
International diffusion of technology: the case of
semi-conductors. Washington, D.C.: Brookings Institution,
1971. 183p. (Studies in the regulation of economic
activity).

3.295 TILTON, John E.

The nature ot firm ownership and the adoption of innov-ations in the electric power industry. Pennsylvania State University, 1973. 21 leaves. Mimeo.

3.296 UNITED NATIONS – Economic Commission for Europe
Aspects ot the diffusion of technology in the United States. A summary of the case study represented by the United States. Geneva: UNECE, 1967. 7 leaves. Mimeo.

3.297 UTECH, Harvey P. and UTECH, Ingrid D.
The communication of innovations between local government departments. A Pilot study for the Office of National R & D Assessment, National Science Foundation. Washington D.C.: National Science Foundation. 65p.

3.298 UTTERBACK, James M.
'Innovation in industry and the diffusion of technology'. Science, 183, 15 February 1974, pp.620–626.

3.299 VON HIPPEL, Eric
Transferring process equipment innovations from user-innovators to equipment manufacturing firms. Cambridge, Mass.: Alfred P. Sloan School of Management, 1977. (revised) 22 leaves. (Working paper 857-76).

3.300 WELLES, John G. and others
The commercial application of missile/space technology. Parts 1 and 2. Denver: Denver Research Institute, 1963.

3.301 WHITE, Mark
The diffusion and adoption of new products with particular reference to textile chemicals. Huddersfield: Department of Marketing Studies, The Polytechnic, 1975. 187p.

3.302 WHITE-HUNT, K.
The application of semiconductor and microcircuit technology (microprocessors) within Scottish manufacturing industry. Stirling: University of Stirling, Technological Economics Research Unit, 1978. 105p. (Discussion paper no.16).

3.303 de BRESSON, Christian
Does Canada fail to innovate? Science Council of Canada, 1979. 22p. + annex. Mimeo.

3.c Specific New Technologies

3.304 ALDRICH, M.J.
Developments in business communication and their impact in
the home. London: Rediffusion Computers, 1981. 13 leaves.
Mimeo.

3.305 ARNOLD, Erik
Information Technology in the home: the failure of PRESTEL.
Paper presented to the EEC FAST Conference, Croydon, January
1982, on "The transition to an information society".
Brighton: SPRU, October 1981. 26p. Mimeo.

3.306 BALL, Robert et al
Transfer of information technology. Stirling: University of
Stirling, Technological Economics Research Unit, 1979. 74p.
(Research Monograph No.3).

3.307 BESANT, C.S.
Computer-aided design and manufacture. Chichester: Ellis
Horwood, 1980. 170p. Pbk. (Ellis Horwood Series in
Engineering Science).

3.308 BESSANT, J.R.
'Microelectronics in the iron and steel industry'. Iron
and Steel International, August 1980, pp.209-213.

3.309 BESSANT, J.R. and DICKSON, K.E.
Issues in the adoption of microelectronics. London: Francis
Pinter, 1982. 148p.

3.310 BONFIGLIOLI, Alberto F.
Ideas for a study on technical innovation related to
materials and energy conservation. Working document for the
Secretariat of the Materials Science and Technology
Committee of the SRC. Brighton: SPRU, 1979. 10 leaves.
Mimeo.

3.311 CANADA - Science Council of Canada/Institute for
         Research on Public Policy
Biotechnology in Canada: promises and concerns. Proceedings
of a workshop prepared by J. Miller, K. Fish, J. Basuk and
Z.P. Zeman. Ottawa: Science Council of Canada, 1980. 62p.

3.312 CANADA - Task Force on Biotechnology
Biotechnology: a development plan for Canada. Report of the
Task Force on Biotechnology to the Minister of State for
Science and Technology. Ottawa: Minister of Supply and
Services, 1981. 51p.

3.313 COMMISSION OF THE EUROPEAN COMMUNITIES
Notes on three French biotechnology reports. A review paper
by Ken Sargeant. Brussels: CEC, 1981. 47p. (FAST)
(XII/1160/81/EN)

3.314 COMPUTER AIDED DESIGN CENTRE
Computer aided design: an appraisal of the present state of
the art. Cambridge: Computer Aided Design Centre, 1978.
28p. Mimeo.

3.315 CORDELL, Arthur J.
Industrial opportunities in the 1980s: trends in
microelectronics and the emergence of biotechnology. First
Global Conference on the Future, Toronto, Ontario, July 23,
1980. Ottawa: Science Council of Canada, 1980. 15p. Mimeo.

3.316 COUNTER INFORMATION SERVICES
The New Technology Report. London: C.I.S., 1979. 40p.

3.317 CURNOW, R.C. and FREEMAN, C.
Product and process change arising from the micro-processor
revolution and some of the economic and social issues.
Keynote address to Institute of Mechanical Engineers, May
1978. 12 leaves. Mimeo.

3.318 CURNOW, Ray
'The future with microelectronics'. New Scientist, Vol.82,
No.1152, 26 April 1979, p.254

3.319 CURNOW, Ray and CURRAN, Susan
The silicon factor: living with the microprocessor. A
National Extension College Discussion Pack linked to the BBC
TV series "The Silicon Factor". Cambridge: National
Extension College, 1979. 72p.

3.320 DARLINGTON, Roger
'Microelectronics - the role of telecommunications'.
Labour Research, Vol.68, No.7, July 1979, pp.160-161.

3.321 FREEMAN, C.
'The economic implications of microelectronics'. In: Agenda
for Britain 1: micro policy choices for the 80's. Edited by
C.D. Cohen. Oxford: Phillip Allen, 1982. p.53-88.

3.322 FURMAN, T.T. (ed)
The use of computers in engineering design. London: English
Universities Press, 1970. 285p.

3.323 GOLDRING, Mary
'Into the electronic Eighties'. The Listener, Vol.101,

No.2612, 24 May 1979, pp.700-1.

3.324 GOLDRING, Mary
'More-from-less (new technology)'. The Listener, Vol.101,
No.2611, 17 May 1980. pp.671-2.

3.325 GRANBERG, Anders
Biological pest control as a technological field: a case
study report. Lund: University of Lund, Research Policy
Institute, 1981. 70p. (Research Policy Studies Discussion
Paper No.139).

3.326 GREAT BRITAIN - Advisory Council for Applied
       Research and Development
Information technology. London: HMSO, 1980. 55p.

3.327 GREAT BRITAIN - Cabinet Office: Information
       Technology Advisory Panel
Report on cable systems. London: HMSO, 1982. 54p.

3.328 GREAT BRITAIN - Civil Service Department, Central
       Computer and Telecommunications Agency
Stand-alone word processors. Report of trials in UK
government typing pools, 1979/80. London: Civil Service
Department, 1980. 34p and Appendices. Mimeo.

3.329 GREAT BRITAIN - Department of Industry
Microelectronics: the new technology. London: HMSO, 1978.
24p.

3.330 GREAT BRITAIN - Department of Industry
Microelectronics: the options. London: HMSO, 1979. 16p.

3.331 GREAT BRITAIN - House of Lords Select Committee on
       the European Communities
New information technologies. Session 1980-81, 27th Report.
With minutes of evidence. London: HMSO, 1981. 150p.

3.332 GREEN, Marjorie
'Automation of foreign exchange and other trading
operations'. The Banker, Vol. 131, No. 622, April 1981, p.
119-123.

3.333 GRIMSDALE, R.L.
'The design and development of microprocessor systems'.
Information Technology, 1978, Conference, p. 449-51.

3.334 GUSTAFSON, Hans
The Lund monitor on technological trends and challenges to
the Third World - the cases of microelectronics and

biotechnology. A project synopsis. Lund, Sweden: Research Policy Institute, University of Lund, 1980. 31p. (Research Policy Studies Discussion Paper no. 137).

3.335 INCOMES DATA SERVICES LTD.
Introducing new office technology. London: IDS, 1981. 17p. (IDS Study 252).

3.336 INSTITUTE FOR INDUSTRIAL RESEARCH AND STANDARDS
Microelectronics - implications for the Irish apparel industry. Report of an investigation conducted on behalf of the National Board for Science and Technology by the Textile Division of the Institute for Industrial Research and Standards. Dublin: IIRS, 1980. 64p. Mimeo.

3.337 JOHNSON, Paul
'Gene-splicing - monster or miracle?' Now, March 20th, 1981, pp.52-8.

3.338 KASTEELE, R.P. Van De
Applications of enzymes. The Hague: Technical and Economic Publishing Office, 1979. 138p. Mimeo.

3.339 LUND, R.T., HALL, L., HORWICH, E. and TAYLOR, D.
Integrated computer-aided manufacturing: social and economic impacts. Cambridge, Mass.: Center for Policy Alternatives, Massachusetts Institute of Technology, 1977. 88p. Mimeo.

3.340 LUND, Robert T. and PINSKY, David A.
Computer-aided materials processing. Report from Seminar, May 1980, Endicott House, Dedham, Massachusetts. Cambridge, Mass.: Center for Policy Alternatives, Massachusetts Institute of Technology, 1980. 94 leaves. Mimeo.

3.341 LUND, Robert T. et al
Microprocessor applications: cases and observations. A report prepared for the Department of Industry by the Massachusetts Institute of Technology. London: HMSO, 1980. 166p. Pbk.

3.342 MARSTRAND, Pauline K.
The state of development of enzyme and other biotechnology in industry. Report of an investigation carried out for the National Board for Science and Technology, Dublin. Brighton: SPRU, May 1980. 23 leaves and Appendices. Mimeo.

3.343 MARSTRAND, Pauline K.
Patterns of change in biotechnology. Brighton: SPRU, June 1981. 83p. (SPRU occasional paper series no.15).

3.344 MARSTRAND, Pauline K.
'Production of microbial protein: a study of the
development and introduction of a new technology'.
Research Policy, Vol.10, No.2, April 1981, pp.148-171.

3.345 MCGLYNN, Daniel R.
Microprocessor technology, architecture and application. New
York, London: Wiley, 1976. 207p.

3.346 MCLEAN, J.M. and RUSH, H.J.
The impact of microelectronics on the U.K.: a suggested
classification and illustrative case studies.   Brighton:
SPRU, 1978. 51p. (SPRU Occasional Paper series No.7).

3.347 MYERS, Sumner
'The space program: a model for technological innovation',
Looking Ahead, 1966, Vol. 14(2), 1-4, 7-8.

3.348 NARANG, Saran A.
Genetic Engineering: the technology and its implications.
Vienna: UNIDO, 1981. 35p. Mimeo.

3.349 NORTHCOTT, Jim
Microprocessors in manufactured products.   London: PSI,
1980. 48p. Pbk. (Report No.590).

3.350 NORTHCOTT, Jim and ROGERS, Petra
Microelectronics in industry: what's happening in Britain.
London: PSI, 1982. 129p. (No. 63).

3.351 RADER, M. and WINGERT, B.
Computer aided design in Great Britain and the Federal
Republic of Germany - current trends and impacts. Karlsruhe:
Kernforschungszentrum Karlsruhe, 1981. 129p. Mimeo.

3.352 RAQUET, John W.E.
The impact of genetic manipulation on the chemical industry.
Guildford: University of Surrey, 1981. 74 leaves. Mimeo.

3.353 RIP, Arie and VAN DER ES, Walter
Biotechnology: its development and context. No imprint. 15
leaves. To appear as Chapter 10 in a book to be published by
Samson (Alphen and Rijn), April 1980. Mimeo.

3.354 RIVARD, Jerome G.
'Microcomputers hit the road.' Special report - automotive
electronics. IEEE Spectrum, Vol.17, No.11, November 1980,
pp.44-47.

3.355 ROSE, A.G. (ed)

Microbial biomass. London: Academic Press, 1979. 459p.
(Economic Microbiology Volume 4).

3.356 ROSENBROCK, H.H.
Computer-aided control system design. London: Academic
Press, 1974. 230p.

3.357 S.R.I. INTERNATIONAL
Seminar on microelectronics. London (1 day), October 1979.
Croydon: SRI International 1979. Various papers.

3.358 SCRIMGEOUR, J.
'CAD/CAM - a challenge and opportunity for Canadian
industry. Advances in computer technology foster widespread
changes in design and manufacturing techniques'.
Engineering Journal, Vol.64, No.4, August 1981, pp.6-12.

3.359 SIM, R.M.
Computer aided manufacture in batch production. A background
paper for the AUEW (TASS) conference on Computer Technology
and Employment, 16 September 1978. Glasgow: National
Engineering Laboratory, 1978. 7p. Mimeo.

3.360 SORGE, Arndt, HARTMANN, Gert, WARNER, Malcolm and
       NICHOLAS, Ian
Microelectronics in the workplace: unity and diversity of
work under CNC in Great Britain and West Germany.   Berlin:
Wissenschaftszentrum Berlin, International Institute of
Management, 1981. 33p. Mimeo.

3.361 STANKIEWICZ, Rikard
The single cell protein as a technological field.   Lund,
Sweden: Research Policy Institute, University of Lund,
1981. 54p. (Research Policy Studies, Discussion Paper,
No.142).

3.362 TATE & LYLE
Biotechnology in the '80's. London: Tate & Lyle Ltd., 1981.
17p.

3.363 THACKRAY, John
'The robots of America'. Management Today, July 1979,
pp.66-69.

3.364 THE UNIVERSITY OF MANCHESTER INSTITUTE OF SCIENCE
       AND TECHNOLOGY
Technological forecasting for downstream processing in
biotechnology: phase 1 - intermediate forecast report.
Manchester: UMIST, 1981. 82 leaves. Mimeo. (European
Economic   Community   FAST   Project   Contract   No.

FST/C/U20/80/UK/H).

3.365 THE WORLD OF BIOTECHNOLOGY
Briefing for top management, 28 October 1980, Selfridge
Hotel, London. Rutland: European Study Conferences Ltd.,
1980. Separately paginated.

3.366 THOMPSON, Gordon B.
Memo from Mercury: information technology is different.
Montreal: Institute for Research on Public Policy, 1979.
62p. (Occasional paper no.10).

3.367 TIDESWELL, M.
Who's afraid of the microchip? London: British Institute
of Management, 1981. 56p. Mimeo.

3.368 VIRGO, Philip
Cashing in on the chips: a policy for exploiting the
semiconductor revolution. London: Conservative Political
Centre, 1979. 32p.

3.369 WILLIAMS, Michael
'The new print technology'. New Society, Vol.48, No.871, 14
June 1979, pp.638-40.

3.370 WILLIAMS, P.W.
'The potential of the microprocessor in library and
information work'. Aslib Proceedings, Vol.31, No.4, April
1979, pp.202-209.

3.371 WILLIAMSON, Robert (ed)
Genetic engineering, 1 & 2. London: Academic Press, 1981. 1
- 167p. 2 - 203p.

3.372 WINSBURY, Rex
'Viewdata in banking: revolutionising the customer
interface?' The Banker, Vol.131, No.622, April 1981,
pp.133-139.

3.373 WISTREICH, J.G.
'The micro-electronics revolution'. New Europe, Vol.6, No.4,
1978. pp.32-38.

3.d   Social Innovation & Technical Change

3.374   CLARK, Peter A.
Action research and organizational change. London: Harper &
Row, 1972. 172p.

3.375   DAVIS, John
Technology for a changing world. Compiled by R. England from
a series of papers by John Davis. London: Intermediate
Technology Pubs. Ltd., 1978. 58p.

3.376   GABOR, Dennis
Innovations: scientific, technological, and social.   London:
O.U.P., 1970. 113p.

3.377   GERSHUNY, J.I.
Innovation and efficiency in the future of the welfare
state. 1st draft. Brighton: SPRU, July 1980. 23p. Mimeo.

3.378   SMITH, Gordon
Social innovation in theory and practice.   Ottawa: Statistics Canada, 1974. 97 leaves. Mimeo.

CHAPTER 4. Economics Of Technical Change

4.a Macro-level Analysis: Technical Change & Economic Growth

4.001 A.U.T.E. ANNUAL CONFERENCE
Causes and consequences of technological change: general information, programme and abstracts. Workshop held at Loughborough University of Technology 2nd-3rd April, 1981. Loughborough: Loughborough University of Technology, 1981. Mimeo.

4.002 AMES, Edward and ROSENBERG, Nathan
'Changing technological leadership and industrial growth.' Economic Journal, 1963, 73(289), 13-31.

4.003 AUSTRALIAN ACADEMY OF SCIENCE - SCIENCE AND INDUSTRY FORUM
The influence of research and development on economic growth. Papers delivered at the 11th Forum meeting, 12 February 1972. Canberra: Australian Academy of Science, 1972. 42p. (Report No. 5).

4.004 BALDWIN, Robert E.
Economic development and growth. New York, Wiley, 1966.

4.005 BARZEL, Yoram
'Productivity in the electric power industry'. Review of Economics and Statistics, 1963, 45(4), 395-408.

4.006 BARZEL, Yoram
'The production function and technical change in the steam-power industry'. Journal of Political Economy, April 1964, 72(2).

4.007 BASKEN, R.C.
'Labour's role in productivity'. Canadian Business Review, Vol. 8, No. 4, Winter 1981. pp.15-16.

4.008 BEAN, Alden S., SCHIFFEL, D.D., and MOGEE, M.E.
The venture capital market and technological innovation. Washington, National Science Foundation, 1974. 55p.

4.009 BECKERMAN, Wilfred
In defence of economic growth. London: Cape, 1974. 287p.

4.010 BHALIA, A.S. (ed)
Technology and employment in industry: a case study
approach. Geneva: I.L.O., 1975. 324p.

4.011 BJERG, Christian and THOGERSEN, John
Technological displacement and compensatory employment.
Aalborg: Aalborg University Center, Institute of
Production, 1981. 24p. Mimeo.

4.012 BLACK, J.
'The technical progress function and the production
function'. Economica, 29, May 1962, 166-170.

4.013 BLACKETT, P.
Technology, industry and economic growth. University of
Southampton, 1966. 13th Fawley Foundation Lecture.

4.014 BLAUG, Mark
'Technical change and Marxian economics'. Kyklos, 1960,
13(4), 495-512.

4.015 BLUNDELL, W.R.C.
'International specialization: a challenge to business'.
Canadian Business Review, Vol.8, No.4, Winter 1981, p.7-9.

4.016 BOND, Floyd A.
Technological Change and Economic Growth. Ann Arbor, Uni-
versity of Michigan Graduate School of Business
Administration, 1965. Michigan Business Papers, 41.

4.017 BONES, Herman P.
'Why has productivity slowed down?' Canadian Business
Review, Vol.8, No.4, Winter 1981, p.10-14.

4.018 BORETSKY, Michel
'The role of innovation: the great productivity debate'.
Challenge, Vol. 23, No.5, November-December 1980, p.9-15.

4.019 BRAENDAARD, A. et al.
The impact of micro-electronics in a small open economy: a
discussion of different approaches. Aalborg: Aalborg
University, Institute for Production, (no date). 47p. Mimeo.
(MIKE Projektet, Smaskrift no. 13).

4.020 BRITTON, N.H. and GILMOUR, J M.
The weakest link: a technological perspective on Canadian
industrial underdevelopment. Ottawa: Science Council of

Canada, 1978. 215p. (Background Study 43. Science Council of Canada).

4.021 BROZEN, Yale
'Economics and changing technology: the economics of automation'. American Economic Review, 47(2), 1957, pp.339-350.

4.022 CAMPOS, Mauricio M.
Adaptations of industrial technology to the factor endowments and other requirements of developing countries. M.A. thesis, Development Economics, University of Sussex, September 1971. 109 leaves.

4.023 CARLSSON, Bo
The content of productivity growth in Swedish manufacturing. Stockholm: IUI, 1980. 21 leaves. Mimeo.

4.024 CARLSSON, Bo
Technical change and productivity in Swedish industry in the post-war period. Stockholm: The Industrial Institute for Economic and Social Research 1980. 40p. Mimeo. (Research Report No.8).

4.025 CARLSSON, Bo
Industrial subsidies in Sweden: macro-economic effects and an international comparison. Paper presented to the European Association for Research in Industrial Economics (EARIE) Conference in Basel, Switzerland, September 16-18, 1981. Draft. Stockholm: Industrial Institute for Economic and Social Research, 1981. 22 leaves. Mimeo.

4.026 CARLSSON, Bo and WALDENSTROM, Erland
Technology, industrial structure and economic growth in Sweden - a 100-year perspective. Stockholm: IUI/IVA, 1979. 32 leaves. Mimeo.

4.027 CARTER, Anne P.
'The economics of technological change'. Scientific American, April 1966, Vol.214(4), 25-31.

4.028 CHARPIE, Robert A.
'Technological innovation and the international economy'. Science Policy News, July 1969, Vol.1(1), 1-2, 4-5.

4.029 CHEMICAL AGE SURVEY
'Survey on West German chemical industry'. Chemical Age, 25 May 1968, Vol.98 (2549), i-xxxii.

4.030 CHESTNUT, H.

Changes in the world induced by technology.   Schenectady,
N.Y.: General Electric, 1972. 23p. (Report No. 72CRD075).

4.031 CHRISTENSEN, L.R., CUMMINGS, D. and JORGENSON, D.W.
'Relative productivity levels, 1947-1973: an international
comparison'.   European Economic Review,   Vol.16,   1981,
p.61-94.

4.032 CLAGE, Ewan and GREENBERG, Leon
'Technological change and employment'.   Monthly Labour
Review, July 1962, Vol.85, 742-746.

4.033 CLARK, John
'A model of embodied technical change and employment'.
Technological Forecasting and Social Change, Vol.16, No.1,
January 1980, p.47-65.

4.034 CLARK, John
Technology and employment - macroeconomic considerations.
Paper prepared for the workshop on: Methodologies to Assess
Impact of Technological Change on Productivity and
Employment, Geneva, 3-5 September 1980. Brighton : SPRU,
September 1980. 32p. Mimeo.

4.035 CLARK, John
Technical change and the prospects for employment - a
suitable case for analysis? Paper presented at the 9th
Triennial Conference on Operational Research I.F.O.R.S.'81,
Hamburg, 20-24 July 1981. Brighton: SPRU, July 1981. 30p.
Mimeo.

4.036 CLARK, John and ROTHWELL, Roy
The employment implications of technological change and R &
D policies. Paper prepared for the Federal Trust for
Education and Research in connection with a study sponsored
by the EEC Commission on: "The employment implications of
policies in the European Community not primarily directed at
the labour market." Brighton: SPRU, October 1980. 43p. +
annotated bibliography. Mimeo.

4.037 CLARK, John, FREEMAN, Christopher and SOETE, Luc
'Long waves, inventions and innovations'. Futures, Vol.13,
No.4, August 1981, p.308-322.

4.038 CLARK, John, FREEMAN, Christopher and SOETE, Luc
'Long waves and technological developments in the 20th
century'. In:- PETZINA, Dietmar and van ROON, Ger (eds)
Konjunkturkrise, Gesellschaft: Wirtschaftliche Wechsellagen
und soziale Entwicklung im 19. und 20. Jahrhundert. Stutt-
gart: Klett-Cotta, 1981. p.132-169.

4.039 CLEMENCE, Richard V. and DOODY, Francis S.
The Schumpeterian system. New York: Augustus M. Kelley,
1966. 117p. (Reprints of Economic Classics).

4.040 COLE, Sam
A model of innovation, employment and income distribution
for a developing economy. Brighton: SPRU, 1979. 15p. Mimeo.

4.041 CONFEDERATION OF BRITISH INDUSTRY
Growth in manufacturing industry. London: CBI, 1981. 36p.
Mimeo.

4.042 CONFERENCE ON ECONOMIC POLICY
De-industrialisation. London, 27 & 28 June 1978. London:
NIESR, 1978. Various papers.

4.043 COOPER, C.M. and CLARK, J.A.
Employment, economics and technology: the impact of techno-
logical change on the labour market. Brighton: Wheatsheaf
Books, 1982. 146p.

4.044 COX, Joan G. and KREIGBAUM, Herbert
Growth, innovation and employment: an Anglo-German
comparison. London: Anglo-German Foundation, 1980. 77p.

4.045 CREEDY, John (ed)
The economics of unemployment in Britain. London: Butter-
worths, 1981. 263p.

4.046 DALY, Donald J.
'Declining productivity growth: an international
perspective'. Canadian Business Review, Vol.8, No.4, Winter
1981, p.2-6.

4.047 DAVID, Paul
Technical choice innovation and economic growth: essay on
American & British experience in the nineteenth century.
London: C.U.P.P. 1975. 334p.

4.048 DAVIES, Duncan and McCARTHY, Callum
Introduction to technological economics. London, Wiley,
1967.

4.049 DELBEKE, Jos
Criticism and classification of real long wave theories.
Louvain, Belgium: Catholic University of Louvain, Centre
for Economic Studies, 1980. 43 leaves. Mimeo.

4.050 DELBEKE, Jos
Towards an endogenous macro-economic interpretation of the

long wave - the case of Belgium, 1830-1980. Paper presented at the 3rd Conference of the Council for European Studies 'Cycles and Periods in Europe: Past and Present' at Washington D.C., April 29-May 1, 1982. Louvain: Katholieke Universiteit, Centrum Voor Economische Studien, 1982. 31 leaves. Mimeo.

4.051 DELL'MOUR, R., FLEISSNER, P. and SINT, P.P.
An input-output approach for the assessment of techno-logical change. Paper presented at the Geneva Workshop on 'Methodologies to Assess the Impact of Technological Change on Productivity and Employment, 1980.' Wien: Oesterreich-ische Akademie Der Wissenschaften, Institut for Sozio-Okonomische Entwicklungsforschung, 1980. No pagination, Mimeo.

4.052 DENISON, Edward F.
'The puzzling setback to productivity growth: the great productivity debate'. Challenge, Vol.23, No.5, November-December 1980, p.3-8.

4.053 DOSI, Giovanni
A simple illustrative model of the impact of increasingly labour-saving technical progress on employment in an open economy. Brighton: SERC, University of Sussex, 1979. 17p. Mimeo.

4.054 ELIASSON, Gunnar
Technical change, employment and growth: experiments on a micro-to-macro model of the Swedish economy. Stockholm: The Industrial Institute for Economic and Social Research, 1979. 31p. Mimeo. (Research Report No.7).

4.055 ELIASSON, Gunnar
Electronics, technical change and total economic perfor-mance. Stockholm: The Industrial Institute for Economic and Social Research, 1980. 45p. Mimeo. (Research Report No. 9).

4.056 FEI, John C.H. and RANIS, Gustav
'Innovation, capital accumulation, and economic development'. American Economic Review, Vol.53(3), June 1963, p.p.283-313.

4.057 FELIX, Fremont
World markets of tomorrow: economic growth, population trends, electricity and energy, quality of life. London: Harper & Row, 1972. 364p.

4.058 FELLNER, William
'Measures of technological progress in the light of recent

growth theories', American Economic Review, Vol. 57(5), 1967, p.p.1073-1098.

4.059 FORRESTER, Jay
'Business structure, economic cycles and national policy'. Futures, June 1976. pp.195-214.

4.060 FRANKEL, Marvin
'The production function in allocation and growth'. American Economic Review, Vol.52(5), 1962, pp.995-1022.

4.061 FREEMAN, C.
'Science policy, technical progress and economic growth'. In: United Nations, Policies and means of promoting technical progress. Papers presented to the fifth meeting of senior economic advisers to ECE governments. New York: UN, 1968.

4.062 FREEMAN, C.
The economics of industrial innovation. Revised edition London and Boston: Francis Pinter and M.I.T. Press, 1982.

4.063 FREEMAN, C.
'The Kondratiev long waves, technical change and unemployment'. In: Organisation for Economic Cooperation and Development, Structural determinants of employment and unemployment. Reports prepared for the experts meeting Paris, 7-11 March 1977. Vol.2. Paris:OECD, 1979. p. 181-196.

4.064 FREEMAN, C. (ed)
'Technical innovation and long waves in world economic development'. Futures, Vol.13, No.4, August 1981, special issue.

4.065 FREEMAN, C., CLARK, John and SOETE, Luc
Unemployment and technical innovation: a study of long waves and economic development. London: Frances Pinter, 1982. 214p.

4.066 FREEMAN, C., KRISTENSEN, Peer Hull and STANKIEWICZ, Rikard
'Policies for technical innovation in the new economic context'. In: Technology policy and industrial development in Scandinavia. Proceedings from a workshop held in Copenhagen, Denmark, May 20-21, 1981, edited by Peer Hull Kristensen & Rikard Stankiewicz. Lund: Research Policy Institute & Roskilde: Institute of Economics & Planning, 1982. p. 21-44.

4.067 GANDER, James P. and others

The relationship of technological change and the demand for and supply of raw materials. Salt Lake City: University of Utah, Bureau of Economic and Business Research, 1976. Vol. 1: Executive Summary. Vol. 2: Final Report. Vol. 3: Analytical Appendices and abstracts.

4.068 GIARNI, Orio and LOUBERGE, Henri
The diminishing returns of technology: an essay on the crisis in economic growth. Oxford: Pergamon Press, 1978. 122p.

4.069 GIBBONS, Michael
'Some implications of low economic growth rates for the development of science and technology in the United Kingdom'. In: Technological Forecasting and Social Change, Vol.17, No.3, July 1980, pp.187-199.

4.070 GIERSCH, Herbert
'Aspects of growth, structural change, and employment - a Schumpeterian perspective'. Weltwirtschaftliches Archiv, Vol. 115, No. 4, 1979, p. 629-651.

4.071 GIERSCH, Herbert (ed)
Emerging technologies: consequences for economic growth, structural change, and employment. Keil Symposium 1981. Tubingen: J.C.B. Mohr (Paul Siebeck), 1982.

4.072 GILMOUR, James M.
Industrialization and technological backwardness: the Canadian industrial dilemma. 43 leaves. Mimeo.

4.073 GLOBERMAN, Steven
Occupational income inequality and technological change: the Canadian experience. For presentation at New York State economics conference, October 30th 1971. 19 leaves. Mimeo.

4.074 GOLD, Bela
Factors stimulating technological progress in Japanese industries: the case of computerization in steel. Cleveland, Ohio: Research Program in Industrial Economics, Case Western Reserve University. No date. 16 leaves. Mimeo. (Working Paper no. 83).

4.075 GOLDSMITH, Maurice (ed)
Technological innovation and the economy. A Science of Science Foundation symposium on 'Technological Innovation and the Growth of the Economy', held at Churchill College, Cambridge, April 11-13, 1969. London, Wiley-Interscience, 1970.

4.076 GORDON, J. Kind (ed)
Canada's role in science and technology development.
Proceedings of a symposium held in Toronto, 10-13 May 1979.
Ottawa: IDRC, 1979. 136p. 9IDRC-141e).

4.077 GOURVITCH, Alexander
Survey of economic theory on technological change and
employment (1940). New York: Augustus M. Kelley, 1966.
252p. (Reprints of Economic Classics).

4.078 GOUVERNEUR, J.
Productivity and factor proportions in less developed
countries: the case of industrial firms in the Congo.
Oxford: Clarendon Press, 1971. 171p.

4.079 GRAHAM, Alan K. and SENGE, Peter M.
A long wave hypothesis of innovation. Cambridge, Mass:
System Dynamics Group, Massachusetts Institute of Technology
No date. 46p. Mimeo (D-3164-1).

4.080 GREAT BRITAIN - Department of Employment
Post-war trends in employment, productivity, output, labour
costs and prices by industry in the United Kingdom.   By
Richard Wragg and James Robertson. London: Dept. of
Employment, 1978. 93p. (Research paper no. 3).

4.081 GREAT BRITAIN - National Economic Development Office
International price competitiveness, non-price factors and
export performance. London: NEDO, 1977. 45p.

4.082 GRILICHES, Zvi
'The sources of measured productivity growth: United States
agriculture,   1940-1960'.   Journal of Political Economy,
August 1963, Vol.71(4), 331-146.

4.083 GRILICHES, Zvi
'Research expenditures, education, and the aggregate
agricultural production function'. American Economic Review,
December 1964, Vol. 54, 962-974.

4.084 GRILICHES, Zvi
'Sources of measured productivity change: capital input'.
American Economic Review, 1966, Vol. 56(2), 50-61.

4.085 GRILICHES, Zvi
R & D and the productivity slowdown. Cambridge, Mass.:
Harvard Institute of Economic Research, Harvard University,
1980. 14 leaves. Mimeo. (Discussion Paper No 749).

4.086 HALTMAIER, Jane

The importance of capital formation in the recent product-
ivity slowdown: a disaggregated approach. San Francisco:
Federal Reserve Bank of San Francisco. 1980. 21 leaves.
(Working paper no. 104).

4.087 HAUSTEIN, Heinz-Dieter and NEUWIRTH, Erich
Long waves in world industrial production, energy,
consumption, innovations, inventions, and patents and their
identification by spectral analysis. Laxenburg: I.I.A.S.A.,
1982. 36p. (Working paper - 82-9).

4.088 HAYASHI, Takeshi
Historical background of technology transfer, transform-
ation, and development in Japan. Tokyo: United Nations
University, 1979. 39p. (Project on Technology Transfer,
Transformation and Development: The Japanese Experience)
HSDRJE-19/UNUP-46).

4.089 HEATHFIELD, David F.
Technical progress, idle capital and the demand for labour.
Paper prepared for the Post-AUTE Conference on Technological
Change held at Loughborough on 2nd and 3rd April, 1981.
Southampton: University of Southampton, 1981. 11p &
Appendices. Mimeo.

4.090 HEERTJE, Arnold
Economics and technical change. London: Weidenfeld &
Nicholson, 1973. 534p.

4.091 HELMER, Olaf
Prospects of technological progress. Santa Monica,
California: Rand Corporation, 1967. P-3646.

4.092 HILL, Christopher T. and UTTERBACK, James M. (eds)
Technological innovation for a dynamic economy. . New York,
London: Pergamon Press, 1979. 344p.

4.093 HSIA, Ronald
Technological change in the industrial growth of Hong Kong.
Mimeo. Paper delivered to the Conference on the Role of
Science and Technology in Economic Development, St. Anton,
August 27 - September 2, 1971.

4.094 HUGH-JONES, E.M.
Economics and technical change. Papers presented to the 1968
Annual General Meeting of the British Association for the
Advancement of Science. Oxford: Basil Blackwell, 1969.
179p.

4.095 INNOVATING INDUSTRY PROJECT

Innovation in industry: a factor for growth. London: Office of Health Economics, 1969. 30p.

4.096   INSTITUTE FOR EMPLOYMENT RESEARCH
Review of the economy and employment, Spring 1982.
University of Warwick, Institute for Employment Research, 1982. 147p.

4.097   INSTITUTE OF WORLD ECONOMICS
Conference on emerging technology: consequences for economic growth, structural change, and employment in advanced open economies. Held at Kiel, June 24-26, 1981. Kiel Institute of World Economics, 1981. Assorted papers. Mimeo.

4.098   INTERNATIONAL LABOUR OFFICE
Higher productivity in manufacturing industries.   Geneva: ILO, 1967. 195p.

4.099   INTRILIGATOR, Michael D.
'Embodied technical change and productivity in the United States, 1929-1958'. Review of Economics and Statistics, February 1965, Vol. 47(1), 65-70.

4.100   JACOBSSON, Staffan
Technical change, employment and technological dependence.
Lund: Research Policy Institute, 1979. 36p. (Discussion paper no. 133).

4.101   JAPAN INDUSTRIAL RESEARCH INSTITUTE
The role of technology in the change of industrial structure. Tokyo: Industrial Research Institute, 1978. 57p.

4.102   JOHNSON, Ellis and STRINER, Herbert E.
Research and development, resources allocation, and economic growth. Bethesda, Md.:   John Hopkins University, 1970. 37p.

4.103   JOHNSON, Harry G.
Technology and economic interdependence. London: MacMillan, for the Trade Policy Research Centre, 1975. 187p.

4.104   JOHNSTON, Robert E.
'Technical progress and innovation'. Oxford Economic Papers, 1966, Vol. 18(2), 158-176.

4.105   JONES, D.I.H.
Technological change, demand and employment.   Leeds: University of Leeds, School of Economic Studies, 1980. 35 leaves. Mimeo. (Discussion Paper No 101).

4.106   JONES, Daniel T.
'Industrial development and economic divergence.' in: E.
Hodges (ed), Economic Divergence and the European Community.
London: Allen and Unwin, 1980. Brighton: SERC, 1979. 30
leaves. Mimeo.

4.107   JORBERG, L(ed.)
Technical change, employment and investment.   Lund:
Department of Economic History, University of Lund, 1982.
240p.

4.108   JORBERG, L. and ROSENBERG, N. (eds)
Technical change, employment and investment. Theme A3 at the
Eighth International Economic History Congress, Budapest
1982. Lund: Department of Economic History, University of
Lund, 1982. 240p. Pbk.

4.109   JORGENSON, Dale W.
'The answer is energy: the great productivity debate'.
Challenge, Vol.23, No.5, November-December 1980, p. 16-25.

4.110   KALDOR, Mary
The baroque arsenal. New York: Hill and Wang, 1981. 294p.

4.111   KANTAROVICH, Leonid V.
'Economic problems of scientific and technical progress'.
Scandinavian Journal of Economics, 1976. p.521-541.

4.112   KATZ, Jorge M.
Production functions, foreign investment and growth: a
study based on the Argentine manufacturing sector 1946-1961.
Amsterdam, London: North-Holland, 1969. 203p. (Contributions to economic analysis no. 58.)

4.113   KENNEDY, Charles
'Technical progress and investment'. Economic Journal, 1961,
Vol. 71(282), 292-299.

4.114   KONDRATIEFF, M.D.
'The long waves in economic life'. Lloyds Bank Review, No.
129, July 1978, p. 41-60.

4.115   KOSHIMURA, Shinzaburo and SCHWARTZ, Jesse G. (eds)
Theory of capital reproduction and accumulation. Kitchener,
Ontario: DPG Publishing Co., 1975. 158p.

4.116   KRISTENSEN, Peer Hull and Levinsen Jorn
The small country squeeze: an evaluation of the scientific
and technological potential of small developed countries in
the emerging international economic division of labor.

Roskilde: Roskilde University Centre with Lund University Research Policy Program, 1978. 337p.

4.117 KRISTENSEN, Thorkill
Inflation and unemployment in the modern society. New York: Praeger, 1981. 169p.

4.118 KUZNETS, Simon S.
Secular movements in production and prices: their nature and their bearing upon cyclical fluctuations. New York: Augustus M. Kelley, 1967. 536p.

4.119 LANCASTER, Kelvin
'Allocation and distribution theory: technological innovation and progress. Change and innovation in the technology of consumption'. American Economic Review, Vol. 56, No. 2, May 1966, p. 14-23.

4.120 LANGRISH, J. and POZNANSKI, Kazimierz
Innovations in the capitalist economy – causes and conditions for their implementation. Manchester, Institute of Advanced Studies, Manchester Polytechnic, 1979. No pagination. Mimeo.

4.121 LEE, Michael
'The economics of innovation'. Proceedings of the Royal Society of Medicine, October 1968, Vol. 61(10), 1051-1054.

4.122 LESSING, Lawrence
'Why the U.S. lags in technology'. Fortune, April 1972, p. 69-72, 146, 149-150.

4.123 LONG, T. Dixon
'Technology and power: Japan catches up'. Japan: the paradox of progress edited by Lewis Austin. New Haven, London: Yale University Press, 1976. Chapter 5. pp.141-164.

4.124 LUCAS, Robert E.
Studies in business-cycle theory. Oxford: Basil Blackwell, 1981. 300p.

4.125 LYDALL, Harold
Technological change and economic growth. University of New England, Department of Economics., 1979. 44 leaves. Mimeo.

4.126 MACHLUP, Fritz
Effects of innovations on the demand for the earnings of productive factors. A report to the National Science Foundation on a survey of the Current State of the Art in the Analysis of the Subject. New York: New York University,

(no date). 99 leaves & appendices. Mimeo.

4.127 MADDISON, A.
Long run dynamics of productivity growth. Paper presented at
the Annual Meeting of the British Association for the
Advancement of Science, held at Bath, 1978. (no imprint) 21
leaves. Mimeo.

4.128 MAGAZINER, Ira C. and HOUT, Thomas M.
Japanese industrial policy: a descriptive account of
postwar developments with case studies of selected
industries. London: Policy Studies Institute, 1980. 90p.
(PSI No. 585).

4.129 MAIER, Harry
Innovation, efficiency and the quantitative and qualitative
demand for labour. Laxenburg, Austria: International
Institute for Applied Systems Analysis, 1980. 25p. Mimeo.

4.130 MANDEL, Ernest
The second slump: a Marxist analysis of recession in the
seventies. Translated by Jon Rothschild. London: NLB, 1977.
212p.

4.131 MANDEL, Ernest
Long waves of capitalist development: the Marxist
interpretation. Based on the Marshall Lectures given at the
University of Cambridge, 1978. Cambridge: CUP, 1980. 151p.
(Studies in Modern Capitalism).

4.132 MANSFIELD, Edwin
The economics of technological change. New York: Norton,
1968.

4.133 MARCHETTI, Cesare
Society as a learning system: discovery, invention and
innovation cycles revisited. Invited paper for a meeting on
'Marketing and Product Innovation Facing Social and
Technological Change', organized by the Italian Association
for Marketing Studies (AISM) in Turin, 18-19, 1980.
Laxenburg, Austria: IIASA, 1980. 30p. Mimeo.

4.134 MASSELL, Benton F.
'Capital formation and technological change in United States
manufacturing', Review of Economics and Statistics,May 1960,
Vol. 42, 182-188.

4.135 MASSELL, Benton F.
Is investment really important? Rand Corporation, 1961.
P-2088.

4.136 MASSELL, Benton F.
Determinants of productivity change in US manufacturing.
Rand Corporation, January 1962.

4.137 MASSELL, Benton F.
'Investment, innovation, and growth', Econometrica, 1962,
Vol. 30(2), 293-252.

4.138 MASSELL, Benton F. and NELSON, Richard R.
Scientific research, economic progress and public policy.
Rand Corporation, January 1961. P-2089.

4.139 MENSCH, Gerhard
Stalemate in technology: innovations overcome the
depression. Cambridge, Mass.: Ballinger, 1979. 241p.

4.140 MENSCH, Gerhard et al
Innovation trends and switching between full- and
under-employment equilibria 1950-1978.   Berlin: International Institute of Management, 1980. 56p.

4.141 MORISHIMA, Michio.
Why has Japan 'succeeded'? Western technology and the
Japanese ethos.   Cambridge: Cambridge University Press,
1982. 207p.

4.142 MOSELEY, Russell
'Technical change & employment in the post-war gas
industry', Omega, Vol. 7, 1979, p. 105-112.

4.143 MacLAURIN, W. Rupert
'Technological progress in some American industries'.
American Economic Review, May 1954, Vol. 44(4), 178-189.

4.144 NADIRI, M. Ishaq
A report on the determinants of research and development
expenditures and their effects on labor productivity of the
firms for the National Science Foundation.   New   York:
National Bureau of Economic Research and Department of
Economics, New York University. No date. 112p. Mimeo.

4.145 NADIRI, M. Ishaq and SCHANKERMAN, M. A.
'Technical change, returns to scale and the productivity
slowdown', American Economic Review, Vol. 71, No. 2, 1981.
p. 314-319.

4.146 NADIRI, M. Ishaq. and SCHANKERMAN, M. A.
The structure of production, technological change and the
rate of growth of total factor productivity in the Bell
system. New York: National Bureau of Economic Research, New

York University. No date. 70 leaves. Mimeo.

4.147 NAU, Henry R.
'A political interpretation of the technology gap dispute'.
Orbis, 15(2), 1971. p.507-527.

4.148 NAU, Henry R.
'The practice or interdependence in the research and
development sector: fast reactor cooperation in Western
Europe', International Organisation, Summer 1972, Vol. 26,
no. 3, p. 499-526.

4.149 NEKOLA, Jiri and VRBA, Josef
Study on the problems of the relation of R & D potential
and the factors of socio-economic development.   Prague:
Czechoslovak Academy of Sciences, 1971. 56p.

4.150 NEKOLA, Jiri and VRBA, Josef
'The optimization of the level of research and development
activity on the basis of the production function',
Ekonomicko-Matematicky Obzor, Vol. 9, No. 1, 1973, p. 51-66.
In Czech. English abstract on p. 66.

4.151 NELSON, Richard R.
'Research on productivity growth and productivity
differences: dead ends and new departures'. Journal of
Economic Literature, Vol. XIX, September 1981, p.1029-1064.

4.152 NELSON, Richard R. and WINTER, Sidney G.
Neoclassical vs. evolutionary theories of economic growth:
critique and prospectus. 45 leaves. Mimeo.

4.153 NELSON, Richard R. and WINTER, Sidney G.
Growth theory from an evolutionary perspective: the
differential productivity puzzle.   New   Haven: Yale
University, Institution for Social and Policy Studies, 1974.
15 leaves. Mimeo. (A Working Paper W4-27)

4.154 NELSON, Richard R. and WINTER, Sidney G. & SCHUETTE,
       Herbert L.
Technical change in an evolutionary model.   Michigan:
University of Michigan, 1973. 64p. Institute of Public
Policy Studies Discussion Paper No. 45.

4.155 NELSON, Richard R. et al
Technology, economic growth and public policy. Washington:
Brookings Institution, 1967.

4.156 NEWLAND, Kathleen
Productivity: the new economic context.   Washington   D.C.:

The Worldwatch Institute, 1982. 47p. (Worldwatch Paper 49).

4.157 NILSSON, Jan-Evert
Is there a cyclical pattern of economic development?   Oslo:
Resource Policy Group, 1980. 26 leaves. Mimeo.

4.158 NORDHAUS, William D.
Invention, growth, and welfare: a theoretical treatment of
technological change. Cambridge, Mass. M.I.T. Press, 1969,
168p.

4.159 ODAGIRI, Hiroyuki
The theory of growth in a corporate economy: management
preference, research and development and economic growth.
Cambridge, CUP, 1981. 220p.

4.160 PASCARELLA, Perry
'Our technological recession: can we again brighten the
dark world?' Across the Board, 16(12), 1979, pp.58-67.

4.161 PAVITT, Keith
'Performance in industrially advanced countries' in: M.
Goldsmith   (ed)   Technological innovation and the economy.
London: Wiley, 1970. Chapter 10, pp.91-107.

4.162 PAVITT, Keith
'Will Europe cross the frontiers of technology?' European
Business, No.35, Autumn 1972, pp.37-44.

4.163 PAVITT, Keith
'Technical change: the   prospects   for   manufacturing
industry'. Futures, Vol.10, No.4, August 1978, pp.283-292.

4.164 PAVITT, Keith
'Technical innovation and industrial development. 1 - The
new   causality'. Futures, Vol.11, No.6, December 1979,
pp.458-470.

4.165 PAVITT, Keith
'Technical innovation and industrial development. 2 - The
dangers of divergence'. Futures, Vol.12, No.1, February
1980, pp.35-44.

4.166 PAVITT, Keith
Technology in British industry: a suitable case for
improvement. Paper prepared for the Conference on Industrial
Policies and Innovation, NIESR, London, 9-10 December, 1980.
Brighton: SPRU, November, 1980. 35 leaves. Mimeo.

4.167 PAVITT, Keith (ed)

Technical innovation and British economic performance.
London: Macmillan, 1980. 353p.

4.168 PAVITT, Keith and SOETE Luc
International differences in economic growth and the
international location of innovation. Paper presented at
Conference on Emerging Technology: consequences for
economic growth, structural change, and employment in
advanced open economies, Institute of World Economics,
University of Kiel, 24-25 June, 1981. (No imprint). 31
leaves. Mimeo.

4.169 PECK, Merton J. and GOTO, Akira
'Technology and economic growth: the case of Japan'.
Research Policy, Vol.10, No.3, July 1981, pp.222-243.

4.170 PRICE, Derek John de Solla
'The acceleration of science - crisis in our technological
civilization'. Product Engineering, 6 March, 1961, 56-59.
Abstracted from a chapter in 'Science since Babylon'.

4.171 PUU, Tonu and WIBE, Soren (eds)
The economics of technological progress. Proceedings of a
conference held by the European Production Study Group in
Umea, Sweden, 23-25 August 1978, including a bibliography on
the subject. London: Macmillan Press, 1980. 336p.

4.172 RAD, P. Someshwar
Factor prices and labour productivity. Ottawa: Economic
Council of Canada, 1981. 77p & appendices. Mimeo.
(Discussion Paper No. 194).

4.173 RANIS, Gustav
'Production functions, market imperfections and economic
development'. Economic Journal, June 1962, 71, 344-354.

4.174 RASMUSSEN, Poul Norregaard
The economics of technological change. Wicksell Lecture,
1975. Stockholm: The Stockholm School of Economics, 1976.
43p.

4.175 RAY, George F.
Innovation and long-term economic growth. Laxenburg,
Austria: IIASA, 1980. 15p. Mimeo. (CP-80-36).

4.176 RAY, George F.
'Innovation as the source of long term economic growth'.
Long Range Planning, Vol.13, No.2, April 1980. pp.9-19.

4.177 RESEK, Robert W.

'Neutrality of technical progress'. Review of Economics and Statistics, February 1963, 45(1), 55-63.

4.178 RICHARDSON, Jacques and PARKS, Ford
'Why Europe lags behind'. Science Journal, August 1968, 4(8), 81-86.

4.179 ROBINSON, Joan
'The production function and the theory of capital'. Review of Economic Studies, 1953/54, 21(55), 81-106.

4.180 ROBINSON, Joan
'Accumulation and the production function'. Economic Journal, 1959, 69(275), 433-442.

4.181 ROSENBERG, Nathan
'Capital goods, technology, and economic growth'. Oxford Economic Papers,1963, 15(3), 217-227.

4.182 ROSENBERG, Nathan
Perspectives on technology. Cambridge: Cambridge U.P., 1976. 353p.

4.183 ROSENBERG, Nathan (ed)
The economics of technological change: selected readings. Harmondsworth: Penguin, 1971. 509p. Pbk.

4.184 ROTHWELL, Roy
'Innovation and structural change in economic development'. Business Graduate,Spring 1982, p.5-7.

4.185 ROTHWELL, Roy
Technology, long waves and economic change. Paper prepared for the Institution of Chemical Engineers "Design 82" Conference, University of Aston, 22-23 September, 1982. Brighton: SPRU, May 1982. Not paginated. Mimeo.

4.186 SAHAL, Devendra
Invention, innovation and productivity growth. New York: New York University, Graduate School of Business Administration, College of Business and Public Administration, 1982. 40 leaves. Mimeo. (Working Paper 82-18).

4.187 SALTER, W.E.G.
Productivity and technical change. 2nd edition. Cambridge: C.U.P., 1966.

4.188 SALVATI, Michele
Science and technology in the new socio-economic context:

technology, long waves and structural unemployment. Paris: OECD, 1977. 8 leaves. Mimeo.

4.189 SATO, Ryuzo and RAMACHANDRAN, Rama
The relationship of technological change and the demand for and supply of raw materials. Final analytical report FY1974-85. 94 leaves. Plus a separate volume, "Classification of abstracts". Mimeo.

4.190 SCHMOOKLER, Jacob
'Technological change and economic theory'. American Economic Review 55(2), 1965, pp.333-341.

4.191 SHAH, Anup and DESAI, Meghnad
'Growth cycles with induced technical change'. The Economic Journal, Vol.91, December 1981, pp.1006-1010.

4.192 SHERIFF, T.D.
The slowdown of productivity growth. London: National Institute of Economic and Social Research, 1980. 18p. Mimeo. (Discussion Paper No.30).

4.193 SIEGEL, Irving H.
'Conditions of American technological progress'. American Economic Review, 44(2), 1954, pp.161-178.

4.194 SOETE, Luc L.G.
'Technical change, catching up and the productivity slowdown'. In: GRANSTRAND, Ove and SIGURDSON, Jon (eds) Technological and industrial policy in China and Europe. Lund: Research Policy Institute, University of Lund, 1981. pp.97-115.

4.195 SOLOW, Robert M.
'Technical change and the aggregate production function'. Review of Economics and Statistics, 39(3), 1957, pp.312-320.

4.196 SOLOW, Robert M.
'Technical progress, capital formation, and economic growth'. American Economic Review, 52, May 1962, pp.76-86.

4.197 SPENCER, D.L.
'An external military presence, technological transfer, and structural change'. Kyklos, 1965, pp.451-474.

4.198 THOGERSEN, John
Some preliminary suggestions on how to generalise the product life-cycle arguments in the "long wave discussion" in a simple macro model, with special regard to employment. Brighton: SPRU, February 1982. 24p. Mimeo.

4.199 TURNER, Roy E.
An interpretation of macro economic technical change.
Revised. Brighton: SPRU, June 1981. 17p. Mimeo.

4.200 UCHINO, Akira
'Technological innovation in postwar Japan'. The
Developing Economies, Vol.7, No.4, December 1969,
pp.406-427.

4.201 UNESCO
The role of science and technology in economic development.
Paris, Unesco, 1970. Science Policy Studies and Documents,
18.

4.202 UNITED NATIONS - Economic Commission for Europe
Consideration of the interrelationship between the economic
growth and trade prospects for the EEC region up to 1990.
Interrelationships between socio-economic development, trade
and technological innovation. Ad hoc meeting on Long-term
Economic Growth and Trade Prospects, Geneva, 28-30 April
1980. Geneva: United Nations, 1980. 10p. Mimeo.

4.203 UNITED NATIONS - Economic Commission for Europe
Factors affecting the impact of science and technology on
long-term economic perspectives. Geneva: United Nations,
1980. 22p. Mimeo.

4.204 UNITED NATIONS - Economic Commission for Europe
Interrelationship between economic growth, structural and
technological change, and trade. Geneva: United Nations,
1980. 28p. Mimeo.

4.205 UNITED NATIONS INDUSTRIAL DEVELOPMENT ORGANIZATION
Restructuring world industry in a period of crisis - the
role of innovation. An analysis of recent developments in
the semiconductor industry. Vienna: UNIDO, 1981. 383p.
(UNIDO/IS.283).

4.206 UNITED STATES - Joint Economic Committee Special
        Study on Economic Change
Research and innovation: developing a dynamic economy.   A
staff study. Washington D.C.:   Joint Economic Committee,
1980. 39p.

4.207 UNITED STATES - Joint Economic Committee, Special
        Study on Economic Change
Productivity: the foundation of growth. Washington D.C.:
USGPO, 1980. 128p. (Special Study on Economic Change,
Vol.10).

4.208 VAN DUIJN, J.J.
'The long wave in economic life'. The Economist, Vol. 125,
No.4, 1977, p.544-576.

4.209 VAN DUIJN, J.J.
On the assessment of the Kondratieff cycle and related
issues: a comment. Delft: Graduate School of Management,
1978. 14 leaves. Mimeo.

4.210 VAN DUIJN, J.J.
Economic policy during a depression. Paper presented at the
Conference 'Managing the unmanageable' Graduate School of
Management at Delft, April 22-23, 1981. Delft, The
Netherlands: Graduate School of Management, 1981. 33
leaves. Mimeo.

4.211 VENABLES, A.J.
The macro-economic implications of a discrete technical
change. Brighton: University of Sussex, Department of
Economics, 1981. 26p. Mimeo. (Discussion Paper No. 81/28).

4.212 VERNON, Raymond
The product cycle hypothesis in a new international
environment. Harvard; Harvard University Press, 1979. 32
leaves. Mimeo.

4.213 WILLIAMS, Bruce
Technology, economic growth and unemployment. Address at the
Conference of the Institute of Ceramics, 15 September 1981.
London: TCC, 1981. 9p. Mimeo.

4.b Firm Behaviour, Competition, Market & Industrial
     Structure

4.214 ABERNATHY, William J.
The productivity dilemma: roadblock to innovation in the
automobile industry. Baltimore/London: John Hopkins
University Press, 1973. 267p.

4.215 ABERNATHY, William J. and ROSENBLOOM, Richard S.
The climate for innovation in industry: the role of
management attitudes and practices in consumer electronics.
Boston, Mass.: Harvard University, Graduate School of
Business Administration, 1982. 36 leaves. Mimeo. (Working
Paper HBS 80-65 Rev. 1/82)

4.216 ABERNATHY, William J. and UTTERBACK, James M.

Innovation and the evolution of technology in the firm.
Revised June, 1976. 37 leaves. Mimeo.

4.217 ANSOFF, Igor H. and STEWART, John M.
'Strategies for a technology-based business'. Harvard Business Review, November/Dec 1967, 45(6), 71-83.

4.218 ANTONELLI, Christiano
'Innovation as a factor shaping industrial structures: the case of small firms', Social Science Information, Vol.18, No. 6, 1979, pp.877-894.

4.219 ARTHUR D. LITTLE INC.
New technology-based firms in the United Kingdom and the Federal Republic of Germany. A report prepared for the Anglo-German Foundation for the Study of Industrial Society. London: The Anglo-German Foundation, 1977. 323p.

4.220 ARTHUR D. LITTLE, INC.
Patterns and problems of technical innovation in American industry. Report to the National Science Foundation. Washington, 1968. PB 181 573.

4.221 BARANSON, Jack
Technology and the multinationals: corporate strategies in a changing world economy. Lexington, Mass.: Lexington Books, D.C. Heath & Co., 1978. 170p.

4.222 BARMA, Tibor
Investment and growth policies in British industrial firms.
NIESR Occasional paper XX. Cambridge, U.P., 1962.

4.223 BRIGHT, James Rieser (ed)
Technological Planning at the Corporate Level. Boston, Harvard University Graduate School of Business Administration, 1962.

4.224 CAIE, Andy and MULDERS, Peter
'The influence of microprocessors on the paper industry'. . Paper, Vol.193, 9 June 1980, p.22, 25, 28 and 34.

4.225 CANADA - Science Council of Canada
Innovation and the structure of Canadian Industry. By Pierre L. Bourgault. Ottawa: Information Canada, 1972. 135p. (Special Study No. 23).

4.226 CANADA - Science Council of Canada
Strategies of development for the Canadian computer industry. Ottawa: Information Canada, 1973. 80p. (Science Council of Canada Report No.21).

4.227 CANADA - Science Council of Canada
The impact of the microelectronics revolution on the
Canadian electronics industry. Proceedings of a Workshop
sponsored by the Science Council of Canada Committee on
Computers and Communication. Ottawa: Science Council of
Canada, 1981. 109p.

4.228 CAREY, Sir Peter
'Telematics: the impact on industry'. The Journal of the
Royal Society of Arts, June 1981, p.401-412.

4.229 CARLSON, Roger D.
Innovation in the firm and the economics of technological
change. Claremont, California, Claremont Graduate School,
1967.

4.230 CARLSSON, Bo, ELIASSON, Gunnar and NADIRI, Ishaq
       (eds)
The importance of technology and the permanence of
structure in industrial growth. Proceedings of a Symposium
at IUI, Stockholm, July 18-19, 1978. Stockholm: The
Industrial Institute for Economic and Social Research, 1978.
237p. (IUI Conference Reports, 1978:2).

4.231 CARTER, Charles F. and WILLIAMS, Bruce R.
'The characteristics of technically progressive firms'.
Journal of Industrial Economics, 1959, Vol.7(2), 87-104.

4.232 CATLING, H. and ROTHWELL, R.
'Automation in textile machinery'. Research Policy, Vol.6,
No.2, April 1977, p.164-176.

4.233 CENTRE FOR THE STUDY OF INDUSTRIAL INNOVATION
Aspects of spin-off: a study of the impact of Concorde and
the advanced passenger train on their supplier-firms.
London: The Centre for the Study of Industrial Innovation,
1971. 44p.

4.234 CHOW, Gregory C.
'Technological change and the demand for computers'.
American Economic Review, 1967, Vol.57(5), 1117-1130.

4.235 COHEN, Benjamin I., KATZ, Jorge and BECK, William T.
Innovation and foreign investment behaviour of the U.S.
pharmaceutical industry. New York: National Bureau of
Economic Research, Inc., 1975. 41p. (Working paper no.101)

4.236 COMANOR, William S.
'Research and competitive product differentiation in the
pharmaceutical industry in the United States'. Economica,

November 1964, Vol.31(124), 372-384.

4.237 COMANOR, William S.
'Research and technical change in the pharmaceutical industry'. _Review of Economic Statistics_, May 1965, Vol.47(2), 182-190.

4.238 CONFEDERATION OF BRITISH INDUSTRY
Innovation and competitiveness in smaller companies. London: CBI, 1979. 41p.

4.239 CONFEDERATION OF BRITISH INDUSTRY
Investment abroad and jobs at home. London: CBI 1980. 39p.

4.240 CONFEDERATION OF BRITISH INDUSTRY - Business Policy
        Department
The future of the Irish clothing industry in the 80's. A report prepared for the Apparel Industries Federation. Dublin: Confederation of Irish Industry, 1980. 53p.

4.241 CONTROL AND AUTOMATION MANUFACTURERS ASSOCIATION -
        Industrial Electronic Systems Section.
The control and automation industry: the current state, the obstacles to growth and the action required. London: The Control and Automation Manufacturers Association, 1976. 11 leaves. Mimeo.

4.242 CORREA, Carlos M.
'Transfer of technology in Latin America: a decade of control'. _Journal of World Trade Law_, Vol.15, No.5, 1981, p.388-409.

4.243 COUNCIL OF EUROPE - Consultative Assembly
Report on the computer industry in Europe: hardware manufacturing. Rapporteur Ian Lloyd, M.P. Strasbourg: Council of Europe, 1971. 65p. In English and French.

4.244 DARRINGTON, Hugh
'Packaging: the next five years'. _Food Manufacturer_, Vol. 56, February 1981, p.25-29.

4.245 DE MELTO, D.P.
Techological change productivity and growth project - micro-level studies. Economic Council of Canada, 1979. 8 leaves. Mimeo.

4.246 DOSI, Giovanni
Industrial adjustment policy: II. Technical change and survival: Europe's semiconductor industry. Brighton: SERC, University of Sussex, 1981. 103p. (Sussex European Papers,

No. 9).

4.247 DOSI, Giovanni
Technology, industrial structures and international
economic performance: an assessment of the state of the art
and some methodological suggestions for future study. Paris:
OECD, 1981. 58p. Mimeo.

4.248 DOSI, Giovanni
Technical change and industrial transformation- the theory
and an application to the semiconductor industry. Brighton:
SPRU, 1982. 374p.

4.249 DUNNING, J.H.
'Profitability, productivity and measures of business
performance'. The Investment Analyst, June 1969. p.22-34.

4.250 DUTTON, Patricia A.
A case study of the G.E.C. - A.E.I. - English Electric
mergers. Coventry: Centre for Industrial Economic and
Business Research, University of Warwick, 1978. 46p. and
Appendices. Mimeo. (General Series Paper No. 81).

4.251 ELSTON, C.D.
'The financing of Japanese industry'. Bank of England
Quarterly, Vol.21, No. 4, December 1981, p. 510-518.

4.252 EVANS, E.
'The British machine tool industry'. Three Banks Review,
June 1964, No. 62, 25-41.

4.253 FRANKO, Lawrence G.
The European multinationals: a renewed challenge to
American and British big business. London: Harper & Row,
1976. 276p.

4.254 FREEMAN, C.
'The plastics industry: a comparative study of research and
innovation'. National Institute Economic Review, No. 25,
November 1963, pp.22-62.

4.255 FREEMAN, C.
Innovation and size of firm. Nathan: Science Policy
Research Centre, Griffith University, 1978. 25p. (Occasional
papers no. 1).

4.256 GIBBONS, N. and WATKINS, D.S.
'Innovation and the small firm'. R & D Management, Vol.
1(1), 1970, pp.10-13.

4.257   GLOBERMAN, Steven
'Market structure and R & D in Canadian manufacturing
industries'. The Quarterly Review of Economics and Business,
Vol.13, (2), Summer 1973. pp. 59-67.

4.258   GOLD, Bela
'Industry growth patterns: theory and empirical results'.
Journal of Industrial Economics, Vol. 13(1), November 1964,
pp.53-73.

4.259   GOLD, Bela
'Productivity, technological change and international
competitiveness'. Technovation, Vol. 1, No. 3, February
1982, p.203-213.

4.260   GOLD, Dela and others
'Long term growth patterns of industries, firms and
products'.   In: Proceedings of the American Statistical
Association, Business and Economic Statistics Section. 1968.
pp.152-161.

4.261   GREAT BRITAIN - Committee of Inquiry on Small Firms
The role of small firms in innovation in the United Kingdom
since 1945. By C. Freeman. London: HMSO, 1971. 43p.
(Research report no.6.).

4.262   GREAT BRITAIN - Printing Industries Sector Working
        Party/Printing and Publishing Industry Training
        Board
Make ready for success: the characteristics of successful
firms in the UK printing industry. London: NEDO, 1981. 23p.

4.263   GREAT BRITAIN - Programmes Analysis Unit
Technology and the garment industry. A   report   by   the
Programmes Analysis Unit on the role of machinery in the
garment industry which has been made available by the
Clothing EDC. London: HMSO, 1971. 188p.

4.264   GREINER, Larry E.
'Evolution and revolution as organizations grow'. Harvard
Business Review, July-August, 1972, p. 37-46.

4.265   HAGGERTY, Patrick E.
'Innovation and the private enterprise system in the United
States', in: National Academy of Engineering, The process
of technological innovation. 1969.

4.266   HAMBERG, D.
'Size of firm, oligopoly and research: the evidence'.
Canadian Journal of Economics and Political Science, 1964,

Vol. 30(1), 62-75.

4.267 HANAN, Mack
'Venturing corporations - think small to stay strong, stick to essentials and spur new growth through business strategies.' Harvard Business Review, May/June, 1976, pp. 139-148.

4.268 HARLOW, Chris
Innovation and productivity under nationalisation: the first thirty years. London: P.E.P. George Allen and Unwin, 1977. 256p.

4.269 HAUSTEIN, Heinz-Dieter
Innovation and industrial strategy. Presented at the Workshop on Innovation Management, June 22-25, 1981. Laxenburg: I.I.A.S.A., 1981. 68p. (Working paper - 81-65).

4.270 HEWITT, Gary
'Research and development performed abroad by U.S. manufacturing multinationals'. Kyklos, Vol.33, No.2, 1980, p. 308-327.

4.271 HILHORST, Jos G.M.
Monopolistic competition, technical progress and income distribution. Rotterdam: U.P., 1965. 152p.

4.272 HIRSCH, Seev
Location of industry and international competitiveness. Oxford, Clarendon, 1967.

4.273 HIRSCH, Seev and ADAR, Zvi
'Firm size and export performance'. World Development, Vol. 2, (7), July 1974. pp. 41-46. 2 copies.

4.274 HOARE & CO.
The heavy electrical plant industry: a survey of the industry's history and structure, an assessment of its present situation and prospects and a description of thirteen leading companies. Prepared by R M A. Wilson and A R. Church. London, 1967.

4.275 HOFFMAN, Kurt and RUSH, Howard
'Microelectronics and the garment industry: not yet a perfect fit'. Bulletin of the I.D.S., Vol.13, no. 2, March 1982. p. 35-41.

4.276 HOLLOMON, J. Herbert
Technical change and American enterprise. Washington, D.C.: National Planning Association, 1974. 44p.

4.277 HOROWITZ, Ira
'Firm size and research activity'. Southern Economic Journal, January 1962, Vol. 28(3), 298-301.

4.278 HOROWITZ, Ira
Decision making and the theory of the firm. New York: Holt, 1970.

4.279 INDUSTRIAL INSTITUTE FOR ECONOMIC AND SOCIAL
       RESEARCH
IUI 40 years, 1939-1979; the firms in the market economy.
Stockholm: IUI, (1980). 150p. (IUI Research Program, 1979-1980).

4.280 INNOVATION IN MARKETING
Innovation in Marketing, 11th Marketing Theory Seminar.
University of Strathclyde, May 1973. Papers presented to the seminar. Strathclyde Univ.: Dept. of Marketing, 1973. Not paginated.

4.281 INSTITUTE OF SCIENCE AND TECHNOLOGY
Electronics in Michigan; a study of the electronic industry in Michigan in its national and international setting. By Frank R. Bacon and Katherine A. Rempp. Ann Arbor, Industrial Development Division, Institute of Science and Technology, University of Michigan, 1967.

4.282 JADLOW, Joseph M.
An empirical study of the relationship between market structure and innovation in therapeutic drug markets.
Stillwater, Oklahoma: College of Business Administration, Oklahoma State University, 1976. 213 leaves. Mimeo.

4.283 JOHANNISSON, Bergt and LINDSTROM, Christian
'Firm size and inventive activity'. Swedish Journal of Economics, No. 4, 1971, p. 427-442.

4.284 JOHNS, B.L. and PAPANDREA, F.G.
Innovation and its financing in the small manufacturing firm. Paper presented at the Eighth Conference of Economics, LaTrobe University, 27-31 August, 1979. Canberra: Bureau of Industrial Economics, 1979. 42 leaves. Mimeo.

4.285 JOHNSON, P.S.
'Firm size and technological change'. Moorgate and Wallstreet, Spring 1970, 5-26.

4.286 JONES, Daniel T.
Catching up with our competitors; the role of industrial policy. First draft. Brighton: SERC, University of Sussex,

1980. 15p. Mimeo. (IAP 80/58).

4.287 JOSEFSSON, Martha and LINSTROM, Bertil
Competition and specialization in the Swedish industry.
Paper prepared for Sixth Conference of European Association
for Research in Industrial Economics (EARIE) held at Ecole
Superieure des Sciences' Economiques et Commerciales
(ESSEL), France, September 9-12, 1979. (No imprint) 25p.
Mimeo.

4.288 JULIEN, Pierre-Andre and LAFRANCE, Christian
On the formalization of 'small is beautiful': the theory of
economies ot scale reconsidered. Quebec: Universite du
Quebec a Trois-Rivieres, 1979. 13 leaves. Mimeo.

4.289 KAMIEN, Morton I. and SCHWARTZ, Nancy L.
Market structure and innovation. Cambridge: Cambridge
University Press, 1982. 241p. (Cambridge Surveys of Economic
Literature).

4.290 KAPLINSKY, Raphael
Computer-aided design: electronics, comparative advantage
and development. London: Frances Pinter, 1982. 144p.

4.291 KUDRLE, Robert T.
Agricultural tractors: a world industry study. Cambridge,
Mass.: Ballinger, 1975. 286p.

4.292 LEA, Sperry and WEBLEY, Simon
Multinational corporations in developed countries: a review
of recent research and policy thinking. London: British
North American Committee, 1973. 77p.

4.293 LEEDS, Simon
'Computerising the clothing industry'. British Clothing
Manufacturer, February, 1981, p. 7-21.

4.294 LEVIN, Richard C.
Toward an empirical model of Schumpeterian competition. New
Haven, Connecticut: Yale University, 1980. 31 leaves.
Mimeo.

4.295 LINK, Albert N.
'Firm size and efficient entrepreneurial activity: a
reformulation of the Schumpeter hypothesis'. Journal of
Political Economy, Vol. 88, No. 4, 1980, p. 771-779.

4.296 MACKINTOSH CONSULTANTS COMPANY LTD.
Microprocessor/minicomputer trends and W. European markets,
1975-1980 - Vol.2: products, suppliers and application.

Fife: Mackintosh, 1976. 484p. Mimeo.

4.297 MACKINTOSH CONSULTANTS COMPANY LTD.
<u>The integrated circuit industry to 1985; a report covering</u>
<u>technology and product trends, economic factors, markets,</u>
<u>strategic influences for integrated circuits and</u>
<u>optoelectronic displays.</u> Luton: Mackintosh International,
1978. 39p.

4.298 MAKOWER, M. (ed)
<u>Transfer processes in technical change – an initial</u>
<u>analysis.</u> Leverhulme Seminar Proceedings. Papers on 23 June
1977. Stirling: Technological Economics Research Unit,
Department of Management Studies, University of Stirling
1977. 84p. Mimeo.

4.299 MANSFIELD, Edwin
'Entry, Gibrat's Law, innovation, and the growth of firms'.
<u>American Economic Review</u>, 1962, Vol. 52(5), 1023-1051.

4.300 MANSFIELD, Edwin
<u>The expenditure of the firm on R & D.</u> New Haven, Conn., Yale
University, 1962. Cowles Foundation Discussion Paper, 36.

4.301 MANSFIELD, Edwin
'Size of firm, market structure, and innovation.' <u>Journal</u>
<u>of Political Economy</u>, 1963, Vol. 71(6), 556-576.

4.302 MANSFIELD, Edwin
'The speed of response of firms to new techniques'.
<u>Quarterly Journal of Economics</u>, 1963, Vol. 77(2), 290-311.

4.303 MANSFIELD, Edwin
'Industrial research and development expenditures:
determinants, prospects, and relation to size of firm and
inventive output.' <u>Journal of Political Economy</u>, 1964, Vol.
72(4), 319-340.

4.304 MANSFIELD, Edwin
<u>Industrial research and technological innovation: an</u>
<u>econometric analysis.</u> New York: Norton, 1968.

4.305 MANSFIELD, Edwin
'Composition of R and D expenditures: relationship to size
of firm, concentration and innovative output'. <u>Review of</u>
<u>Economics and Statistics</u>, Vol LXIII, No. 4, 1981, p.
610-615.

4.306 MANSFIELD, Edwin
<u>Studies of the relationship between international</u>

technology transfer and R and D expenditure by U.S. firms.
Philadelphia: University of Pennsylvania. No date 103
leaves. Mimeo.

4.307 MANSFIELD, Edwin and ROMEO, Athony
'Technology transfer to overseas subsidiaries by U.S.- based
firms'. Quarterly Journal of Economics, December 1980, p.
737-750.

4.308 MANSFIELD, Edwin et al
Research and innovation in the modern corporation. London:
MacMillan, 1972. 239p.

4.309 MANSFIELD, Edwin, TEECE, David and ROMEO, Anthony
'Overseas research and development by US-based firms'.
Economica, Vol. 46, p. 187-196.

4.310 MARKHAM, Jesse W.
'Market structure, business conduct, and innovation',
American Economic Review, 1965, Vol. 55(2), 323-332.

4.311 MARKHAM, Jesse W. and PAPANEK, Gustav F.
Industrial organization and economic development: in honour
of E.S. Mason. Edited by Jesse W. Markham. and Gustav F.
Papanek. New York: Houghton Mifflin Co., 1970. 422p.

4.312 MARKS, C.P.
'The impact on the user of the microcomputer revolution'.
Management Services in Government, Vol.33, No.4, November
1978, pp.177-184.

4.313 MAXWELL, Philip and TEUBEL, Morris.
Capacity-stretching technical change: some empirical and
theoretical aspects. Buenos Aires: UNECLA, 1980. 85p.
Mimeo. (Research Programme on Scientific and Technological
Development in Latin America, Working Paper No. 36)

4.314 MILLER, Arthur Selwyn.
'Corporate gigantism and 'technological imperatives'',
Journal of Public Law, 1969, Vol. 18(2), 256-310. Program of
Policy Studies in Science and Technology, Reprint 6.

4.315 MILLER, Stephen M. and ROMEO, Anthony A.
'Alternative goals and uncertainty in the theory of the
firm', Southern Economic Journal, Vol. 46, No. 1, July 1979,
p. 189-205.

4.316 MORGAN, Alun and BLANPAIN, Roger
The industrial relations and employment impacts of
multinational enterprises: an inquiry into the issues.

Paris: OECD, 1977. 42p.

4.317 MOUNTFIELD, P. R. and UNWIN, D. J. & GUY, K.
Processes of change in the footwear industry of the East
Midlands. Leicester: University of Leicester, Department of
Geography, 1982. 340p & Appendices. Mimeo.

4.318 MOWERY, David. and ROSENBERG, Nathan.
The influence of market demand upon innovation: a critical
review of some recent empirical studies. Stanford California: Stanford University, no date. 73 leaves. Mimeo.

4.319 MacLAURIN, W.Rupert
'The process of technological innovation: the launching of
a new scientific industry'. American Economic Review, 1950,
Vol. 4(1), 90-112.

4.320 McGEEHAN, Joy M.
'Competitiveness: a survey of recent literature'. The
Economic Journal, June 1968, Vol. 78(310), 243-262.

4.321 McLEAN, J. Michael
The impact of the microelectronics industry on the
structure of the Canadian economy. Montreal: Institute for
Research on Public Policy, 1979. 50p. (Occasional paper no.
8)

4.322 McLEAN, Mick
Science and technology in the new economic context: report
of the Electronics Sector Survey. Brighton: SPRU. 1978. 70
leaves. Mimeo.

4.323 NEEDHAM, Douglas
'Market structure and firms R & D behaviour', Journal of
Industrial Economics, Vol. XXIII, no. 4, June 1975, p.
241-255.

4.324 NELSON, Richard R.
Issues and suggestions for the study of industrial
organistion in a regime of rapid technical change. Yale
University, Economic Growth Centre, 1971. 35 leaves. Mimeo.

4.325 NELSON, Richard R.
R & D knowledge and externalities: an approach to the
puzzle of disparate productivity growth rates among
manufacturing industries. New Haven, Conn.: Yale
University, Institution for Social and Policy Studies, 1978.
Revised edition. 30 leaves. Mimeo.

4.326 NELSON, Richard R.

Parsimony, responsiveness, and innovativeness as virtues of private enterprise: an exegesis of tangled doctrine. New Haven, Connecticut: Yale University, Institution for Social and Policy Studies, 1980. 48p. Mimeo. Working Paper 828

4.327 NELSON, Richard R. and WINTER, Sidney G.
Dynamic competition and technical progress. New Haven: Yale University, 1975 Revised February 1976. 68 leaves. Institution for Social and Policy Studies, working paper no. 763.

4.328 NELSON, Richard R. and WINTER, Sidney G.
Firm and industry response to changed market conditions: an evolutionary approach. New Haven: Institution for Social and Policy Studies, Yale University, 1977. 44 leaves. Mimeo.

4.329 NELSON, Richard R. and WINTER, Sidney G.
'Forces generating and limiting concentration under Schumpeterian competition', Bell Journal of Economics, Vol. 9, No. 2, Autumn 1978, p. 524-548.

4.330 NEUMANN, M., BOBEL, I. and HAID, A.
Market structure and the labour market in West German industries: a contribution to interpreting the structure-performance relationship. Paper presented to the 6th Conference on Industrial Structure (EARIE), 9-11 September, 1979, Paris. Nurnberg: Volkswirtschaftliches Institut der Universitat Erlangen-Nurnberg, 1979. 18p. Mimeo.

4.331 NORTHCOTT, Jim and ROGERS, P.
Microelectronics in industry: what's happening in Britain. London: PSI, 1982. 129p. (No.63).

4.332 NORTHCOTT, Jim with ROGERS, P. and ZEILINGER, A.
Microelectronics in industry: advantages and problems. London: Policy Studies Institute, 1981. 62p.

4.333 NUTTER, G. Warren
'Monopoly, bigness and progress'. Journal of Political Economy, December 1956, 64(6), 520-527.

4.334 NYSTROM, Harry
Company strategies for research and development. Uppsala: The Agricultural College of Sweden, 1977. 56p. (Rapport fran Institutionen for Ekonomi och Statistik, nr. 107).

4.335 ORR, John L.
'A technological strategy for industrial progress in a competitive world'. Science Forum, 1968, 1(4), 3-7.

4.336 PANIC, M. (ed)
The UK and West German manufacturing industry 1954-72: a
comparison of structure and performance.   London: NEDO,
1976. 151p. (NEDO Monograph 5).

4.337 PARKER, E.F.
'British chemical industry in the 1980's - a Delphi method
profile'. Chemistry and Industry, 31 January 1970, 138-145.

4.338 PARKER, J.E.S.
The economics of innovation: the national and multinational
enterprise in technological change. London, Longman, 1974.
297p.

4.339 PAVITT, Keith
'Technological innovation in European industry: the need
for a world perspective'. Long Range Planning, December
1969, pp.8-13.

4.340 PAVITT, Keith
R & D, innovation and exports in the British electrical
industry: some trends and international comparisons.
Brighton: SPRU, March, 1979. 10 leaves. Mimeo.

4.341 PENROSE, Edith T.
'Foreign investment and the growth of the firm'. Economic
Journal, June 1956, pp.220-235.

4.342 PENROSE, Edith T.
The theory of the growth of the firm. 2nd ed. Oxford: Basil
Blackwell, 1980. 272p.

4.343 PHILIPS INDUSTRIES
Innovation in a multinational industrial company. (No imprint). 1979. 11p. + charts.

4.344 PHILLIPS, Almarin
'Concentration, scale and technological change in selected
manufacturing   industries,   1899-1939'.   Journal of
Industrial Economics, 4(3), 1956, 179-193.

4.345 PHILLIPS, Almarin
Technology and market structure: a study of the aircraft
industry. Toronto, London: Lexington, 1973. 233p.

4.346 PISANO, Gary and SOETE, Luc
Diversification of innovation, firm size and R & D.
Brighton: SPRU, December, 1981. 15 leaves + tables. Mimeo.

4.347 RAQUET, John W.E.

The impact of genetic manipulation on the chemical industry.
Guildford: University of Surrey, 1981. 74 leaves. Mimeo.

4.348 RAY, George F.
New technology and enterprise decisions. Paper presented at
the International Conference on Industrial Economics,
Budapest, 15-17 April, 1970. pp.349-367.

4.349 REEKIE, W.D.
The economics of innovation with special reference to the
pharmaceutical industry. Phd. Thesis. Strathclyde, 1971. 2v.

4.350 RENDEIRO, Joao Oliveira and SHEPHERD, Geoffrey
Technical change and competitiveness in the scientific and
industrial instrument industry of the EEC. Report prepared
for the Commission of the European Communities. Brighton:
University of Sussex, SERC, 1981. 200p. Mimeo.

4.351 RHEE, Yung W. and WESTPHAL, Larry E.
A micro, econometric investigation of the impact of
industrial policy on technology choice. Washington D.C.:
The World Bank, 1976. 49 leaves. Mimeo. (Development
Research Center Discussion Papers, No.19).

4.352 ROBSON, M.
U.K. activities in underwater engineering- the Vickers
story and a bit more. Brighton: SPRU, January 1983. Mimeo.

4.353 ROSENBERG, Nathan
'Technological change in the machine tool industry
1840-1910'. Journal of Economic History, 1963, 23(4),
414-446.

4.354 ROSENBERG, Nathan and STEINMUELLER, W. Edward
'The economic implications of the VLSI revolution'. Futures,
Vol.12, No.5, October 1980, pp.358-369.

4.355 ROTHWELL, Roy
'Small and medium sized manufacturing firms and
technological innovation'. Management Decision 16, (6),
1978, pp.362-370.

4.356 ROTHWELL, Roy
'The relationship between technical change and economic
performance in mechanical engineering: some evidence'. In:
BAKER, Michael J. (ed) Industrial innovation: technology,
policy, diffusion. London: Macmillan Press, 1979. pp.36-59.

4.357 ROTHWELL, Roy
'Firm size and innovation: a case of dynamic

complementarity', <u>Journal of General Management</u>, Spring Issue, 1983.

4.358 ROTHWELL, Roy and GARDINER, Paul
<u>The role of design in competitiveness.</u> Paper prepared for: International Conference on Design Policy, Royal College of Art, London, 20-23 July 1982. Brigton: SPRU, June 1982. 28p. Mimeo.

4.359 ROTHWELL, Roy and ZEGVELD, Walter
'Small manufacturing enterprises in the innovation process: advantages and disadvantages'. <u>Planned Innovation</u>, Vol.2, No.9, September 1979, pp.317-319.

4.360 ROTHWELL, Roy and ZEGVELD, Walter
'The role of small manufacturing enterprises in innovation: an overview of recent research'. <u>Planned Innovation</u> Vol.2, No.1, January 1979. pp.3-5.

4.361 ROTHWELL, Roy and ZEGVELD, Walter
<u>Innovation and the small and medium sized firm: their role in employment and in economic change.</u> London: Frances Pinter, 1982. 268p.

4.362 ROY, R., WALKER, D. and WALSH, V.
<u>Design, innovation and competitiveness in British manufacturing industry.</u> Milton Keynes: Design Innovation Group, Faculty of Technology, The Open University, 1981, 10p. Mimeo.

4.363 SANDERSON, Michael
'Research and the firm in British industry, 1919-39'. <u>Science Studies</u> Vol.2, 1972, pp.107-151.

4.364 SANDKULL, Bengt
<u>Innovative behaviour of organizations: the case of new products.</u> Lund: Studentlitteratur, 1970. 165p. Pbk.

4.365 SCHERER, F.M.
<u>Inter-industry technology flows and productivity growth .</u> Chicago: Northwestern University, (no date), 38 leaves. Mimeo. Added to stock, November 1981.

4.366 SCHERER, F.M.
'Firm size, market structure, opportunity, and the output of patented inventions'. <u>American Economic Review</u>, 55(5), Part 1, December 1965, 1097-1123.

4.367 SCHERER, F.M.
'Size of firm, oligopoly and research: a comment'.

Canadian Journal of Economics and Political Science, 31(2), May 1965, 256-266.

4.368 SCHERER, F.M.
Industrial market structure and economic performance.
Chicago: Rand McNally & Co., 1970. 576p.

4.369 SCIBERRAS, Ed.
Multinational electronics companies and national economic policies. Greenwich, Conn.: Jai P., 1977. 328p.

4.370 SCIBERRAS, Ed., SWORDS-ISHERWOOD, Nuala and SENKER, P.
The theory of the firm, technical change, manpower and competitiveness. 2nd edition. September 1977. 56 leaves. Mimeo.

4.371 SCOTT-KEMMIS, D
'Industrial learning- evidence from improvement curves', Paper presented to the Second conference on technology and industry policy in China and Europe, University of Sussex, September, 1982.

4.372 SHARPE, William F.
The economics of computers. New York, London: Columbia U.P., 1969. 571p.

4.373 SHAW, R.W.
'New entry and the competitive process in the U.K. fertiliser industry'. Scottish Journal of Political Economy, Vol.27, No.1, February 1980, pp.1-16.

4.374 SHAW, R.W. and SHAW, S.A.
'Patent expiry and competition in polyester fibres'. Scottish Journal of Political Economy, Vol.24, No.2, June 1977, pp.117-132.

4.375 SHEPHERD, Geoffrey
Public and private strategies for survival in the textile and clothing industries of Western Europe and the United States. Brighton: SERC, University of Sussex, September 1980. 58p. Mimeo.

4.376 SHIMWELL, Charles and WHITEHUNT, Keith
The application and utilisation of computer and microelectronic technology within manufacturing and business systems with particular reference to the steel industry.
Stirling: Technological Economics Research Unit, Department of Management Science and Technology Studies. University of Stirling, 1979. 92p. Mimeo. (Discussion Paper No.22).

4.377 SHIRLEY INSTITUTE
E.E.C. textile trade: proposal for a technological forecast
of the likely future situation of the textile industries of
the E.E.C. London: Textile Research Conference, no date.
13p.

4.378 SHIRLEY INSTITUTE
The U.K. textile industry in 1980: a technical forecast
made using a modification of the Delphi technique. By Ms. P.
Rodgers. Manchester: Shirley Institute, 1972. 475p.

4.379 SHISHIDO, Toshio
Technology transfer — from the standpoint of foreign direct
investment by Japanese corporations. Tokyo: Nikko Research
Center, 1976. 19 leaves. Mimeo.

4.380 SHREVE, R. Norris
The chemical process industries. New York: McGraw—Hill,
1965. 1004p. (McGraw—Hill Series in Chemical Engineering.
Sidney D. Kirkpatrick, Consulting Editor).

4.381 SHRIEVES, Ronald E.
'Market structure and innovation: a new perspective'.
The Journal of Industrial Economics, Vol.XXVI, No.4, June
1978, pp.329—347.

4.382 SIMMONDS, W.H.C.
Industrial behaviour patterns and their significance to
B.C.'s industrialization. Ottawa: National Research Council
of Canada, 1973. 12 leaves. Mimeo.

4.383 SIMMONDS, W.H.C.
Industries reclassified: today's behaviour shapes
tomorrow's results. Paper presented at the Commercial
Development Association/Chemical Market Research Association
Conference, St. Louis, Missouri, November 26—28, 1973.
Ottawa: National Research Council of Canada, 1973. 14
leaves. Mimeo.

4.384 SIMMONDS, W.H.C.
'Toward   an   analytical   industry   classification'.
Technological Forecasting and Social Change, Vol.4, 1973,
pp.375—385.

4.385 SIMMONDS, W.H.C.
'Industrial behaviour patterns: a new dimension for
planners'. Futures, August 1975, pp.284-292.

4.386 SIMMONDS, W.H.C.
'Patterns of industry behaviour and what they tell us'.

Chemtech, July 1975, pp.416-420.

4.387   SIX COUNTRIES PROGRAMME ON ASPECTS OF GOVERNMENT
        POLICIES TOWARDS TECHNOLOGICAL INNOVATION IN
        INDUSTRY
New entrepreneurship and the smaller innovative firm. Papers
presented at the Workshop on Entrepreneurship and the
Smaller Innovative Firm, Limerick, June 9-10, 1980. Delft,
Netherlands: Secretariat Six Countries Programme, 1980.
76p. Mimeo.

4.388   SMITH, Donald N.
'Why companies balk at technology transfers'. Columbia
Journal of World Business, II(3), May-June 1967, pp.45-53.

4.389   SMITH, Donald N.
Technological change in Michigan's tool and die industry.
Ann Arbor, Michigan: University of Michigan. Institute of
Science and Technology, 1968. 140p. Pbk.

4.390   SOETE, Luc L.G.
'Firm size and inventive activity: the evidence
reconsidered'. European Economic Review,   Vol.12,   1979,
pp.319-340.

4.391   SOETE, Luc L.G.
'Size of firm, oligopoly and research: a reappraisal'.
From: Reseaux, No.35-36, 1979, pp.99-125, Special Issue on
Science Policy and Social Aspects of Science.

4.392   SOLOW, Robert M.
'Investment and technical progress'. In: ARNOW, K.
Mathematical models in the Social Sciences.   Stanford
University Press, 1960, Chapter 7.

4.393   SOMA, John T.
The computer industry: an economic/legal analysis of its
technology and growth. Lexington: Lexington Books, 1976.
219p.

4.394   SORENSON, Ralph Z.
'U.S. marketers can learn from European innovators'.
Harvard Business Review, September/October, 1972, pp.89-99.

4.395   STONEMAN, P.S.
Merger and technological progressiveness: the case of the
British computer industry. Conventry: University of
Warwick, 1975. 22p. (Warwick Economic Research Papers
No.79).

4.396 STOUT, H.P. and BRADBURY, F.R.
Technology transfer practice of small and medium sized
enterprises. Stirling TERU, University of Stirling, 1979.
31p. (Research Monograph No.2).

4.397 STRASSMAN, W. Paul
'Interrelated industries and the rate of technological
change'. Review of Economic Studies 27(72), October 1959,
pp.16-22.

4.398 SURREY, John and WALKER, William
The European power plant industry: structural responses to
international market pressures. Brighton: University of
Sussex, Sussex European Research Centre, 1981. 90p.
(Industrial Adjustment and Policy: III. Sussex European
papers No.12).

4.399 TEELING-SMITH, G.
'Innovative activity in the pharmaceutical industry'.
Chemistry and Industry, 2 June 1973, pp.502-4. Paper
presented at a conference, London, 3 May 1973, organised by
NEDO.

4.400 TEELING-SMITH, George (ed)
Economics and innovation in the pharmaceutical industry.
Symposium held at the Imperial College of Science and
Technology by the Office of Health Economics in 1968-69.
London: Office of Health Economics, 1969.

4.401 TERLECKYJ, Nestor E.
Effects of R & D on the productivity growth of industries:
an exploratory study. Washington, D.C.: National Planning
Association, 1974. 63p.

4.402 TEUBAL, Morris
The R & D performance through time of high-technology
firms: methodology and an illustration. Draft. Jerusalem:
The Maurice Falk Institute for Economic Research in Israel,
1981. 27p. Mimeo. (Discussion paper no.814).

4.403 TEUBAL, Morris, ARNON, N. and TRACHTENBERG, M.
'Performance in innovation in the Israeli electronics
industry: a case study of biomedical electronics
instrumentation'. Research Policy, Vol.5, No.4, October
1976, pp.354-379.

4.404 TILTON, John E.
Firm size and innovative activity in the semiconductor
industry. September 1971. 28p. Mimeo.

4.405 TOWNSEND, Harry
Scale, innovation, merger and monopoly. Oxford: Pergamon,
1968. 106p.

4.406 UNITED STATES - Department of Labor
Technological change and manpower trends in five
industries: pulp and paper; hydraulic cement; steel;
aircraft and missiles; wholesale trade. Washington D.C.:
USGPO, 1975. 57p. Mimeo. (Bulletin 1856).

4.407 UNITED STATES - General Accounting Office
Consistent criteria are needed to assess small business
innovation initiatives. Washington D.C.: USGAO, 1981. 86p.
Mimeo.

4.408 UNITED STATES - General Accounting Office
Small businesses are more active as inventors than as
innovators in the innovation process. Report to the
Chairman, Committee on Small Business, United States House
of Representatives. Washington D.C.: USGAO, 1981. 20p.
Mimeo.

4.409 UNITED STATES - Office of Technology Assessment
U.S. industrial competitiveness: a comparison of steel,
electronics and automobiles. Washington D.C.: U.S. O.T.A.,
1981. 206p.

4.410 UTTERBACK, James M. and ABERNATHY, William J.
A test or a conceptual model linking stages in a firm's
process and product innovation. Boston, Mass.: Harvard
University Graduate School of Business Administration, 1974.
21p. Mimeo.

4.411 UTTERBACK, James M., BOLLINGER, Lynn and HOPE,
      Katherine
Technology and industrial innovation in Sweden: a review of
literature and hypotheses on new technology-based firms.
Cambridge, Mass.: Center for Policy Alternatives,
Massachusetts Institute of Technology, 1981. 33 leaves.
Mimeo.

4.412 VENABLES, A.J.
Monopolistic competition and the possible losses from trade.
Brighton: University of Sussex, 1980. 23p. Mimeo. (Economic
Seminar Paper Series 1980/1).

4.413 VERNON, Raymond and DAVIDSON, W.H.
Foreign production of technology-intensive products by U.S.
based multinational enterprises. Boston: Harvard University
Graduate School of Business Administration, 1979. 148p.

Working paper.

4.414 WAGNER, Karin
'Competition and productivity: a study of the metal can industry in Britain, Germany and the United States'. The Journal of Industrial Economics, Vol.XXIX, No.1, September 1980, pp.17-35.

4.415 WALKER, W.B.
Technical change and economic performance in the UK mechanical engineering industry: a review of the literature. Report to Programmes Analysis Unit. 30 June 1976. 56 leaves. Mimeo.

4.416 WALSH, V., TOWNSEND, J., ACHILLADELIS, B. and
        FREEMAN, C.
Trends in invention and innovation in the chemical industry. Report to the SSRC. Brighton: SPRU, July 1979. Not paginated. Mimeo.

4.417 WEBBINK, Douglas
The semi-conductor industry: a survey of structure, conduct and performance. Staff report to the Federal Trade Commission of the U.S. January 1977. 195p.

4.418 WEISS, Frank D.
The structure of Germany's international competitiveness - an appraisal. Kiel: The Kiel Institute for World Economics, 1981. 62 leaves. Mimeo.

4.419 WHISTON, Thomas G.
U.S. regulatory control and innovatory response in the automobile industry - a view from Europe. Brighton: SPRU, June 1980. 185 leaves. Mimeo.

4.420 WILLIAMSON, Oliver E.
'Innovation and market structure'. Journal of Political Economy, 73(1), 1965, pp.67-73.

4.c International Trade, Technical Change And Innovation

4.421 AHO, Michael C. and BOWEN, Harry P.
Why U.S. industry is losing and may continue to lose its competitive edge. Washington DC: U.S. Department of Labour, 1981. 4 leaves. Mimeo

4.422 AHO, Michael C. and ROSEN, Howard F.

<u>Trends in technology-intensive trade: with special</u>
<u>reference to U.S. competitiveness.</u> Washington DC: U.S.
Department of Labour, 1980, 59p. Mimeo. (Economic Discussion
Paper 9).

4.423 AISLABIE, C.J.
<u>Product quality and export income.</u> Department of Economics
Seminar, 28 May 1982. Australia, N.S.W.: University of
Newcastle, Department of Economics, 1982, 31p. Mimeo.

4.424 ALLEN, Thomas J.
'The international technological gatekeeper'. <u>Technology
Review</u>, March 1971, p.37-44.

4.425 ANDERSEN, Esben Sloth, DALUM Bent and VILLUMSEN,
       Gert
<u>The home market's importance for the development of</u>
<u>technology and the export specialization of manufacturing</u>
<u>industry.</u> First draft. Aalborg: University of Aalborg,
Institute of Production, Project IKE, 1980. 59 leaves.
Mimeo.

4.426 BALASSA, Bela
'Tariff protection in industrial countries: an evaluation'.
<u>Journal of Political Economy</u>, December 1965, 73(6), 573-594.

4.427 BARANSON, Jack
<u>Sources of Japan's international competitiveness in the</u>
<u>consumer electronics industry.</u> Washington DC: Developing
World Industry and Technology Inc., 1980. 175p. Mimeo.

4.428 BARANSON, Jack
<u>The Japanese challenge to U.S. industry.</u> Lexington, Mass.:
Lexington Books, 1981. 188p.

4.429 BASEVI, Giorgio
'The United States tariff structure: estimates of effective
rates of protection of United States industries and
industrial labor'. <u>Review of Economics and Statistics</u>, May
1966, 48(2), 147-160.

4.430 BEHRMAN J.N.
'Foreign investment and the transfer of knowledge and
skills', in: MIKESELL, Raymond <u>U.S. private and government</u>
<u>investment abroad.</u> University of Oregon Press, 1962.

4.431 BEHRMAN, J.N.
'Advantages and disadvantages of foreign licensing'.
<u>Patent, Trademark and Copyright Journal of Research and</u>
<u>Education</u>, 1958. 137-158.

4.432 BEHRMAN, J.N. and SCHMIDT, W.E.
'Royalty provisions in foreign licencing contracts'. Idea,
February 1959, pp. 272-

4.433 BEHRMAN, J.N. and WALLENDER, H.W.
Transfers of manufacturing technology within multinational
enterprises. Cambridge, Mass.: Ballinger Publishing Co.,
1976. 308p.

4.434 BESAROVIC, Vesna
Legal aspects of the transfer of technology in modern
society. Tokyo: United Nations University, 1980. 18p.
Mimeo. (HSDRSCA-30/UNUP-175)

4.435 BLACKURST, Richard, MARIAN, Nicolas and TUMLIR, Jan
Adjustment, trade and growth in developed and developing
countries. Geneva: General Agreement on Tariffs & Trade,
1978. 98p. (GATT Studies in International Trade).

4.436 BOON, Gerard K.
Technology transfer in fibres, textile and apparel. Alphen
aan den Rijn, the Netherlands: Sijthoff & Noordhoff, 1981.
581p.

4.437 BROOKS, Harvey
'What's happening to the U.S. lead in technology?' Harvard
Business Review, May/June 1972, p.110-118.

4.438 CANADA - Science Council of Canada
Multinationals and industrial strategy: the role of world
product mandates. A statement prepared by the Science
Council Working Group on Industrial Policies. Ottawa:
Science Council of Canada, 1980. 77p.

4.439 CANADA - Science Council of Canada
The adoption of foreign technology by Canadian industry.
Proceedings of a Workshop sponsored by the Science Council
of Canada Industrial Policies Committee. Ottawa: The
Council, 1981. p.152. (Proceedings P81/2).

4.440 CARRERE, Maximo Halty
An experiment in international transfer of technology : a
pilot project for Latin America. Washington, D.C.:
Organisation of American States, 1972. 15 leaves.

4.441 CARRICK, R J.
East-West technology transfer in perspective. Berkeley:
Institute ot International Studies, University of
California, 1978. 93p. (Policy Papers in International
Affairs - Number Nine).

4.442 CHAPMAN, J.W.M.
The transfer of German underwater weapons technology to Japan, 1919-1976. European Association of Japanese Studies, 1st. International Conference, Zurich, Sept.1976. No imprint. 13p. Mimeo.

4.443 CIKATO, Manfredo A.
International transfer of technology. Montevideo, Uruguay: Barbet & Cikto, Patent Lawyers, 1974. 39p.

4.444 CLARK, Norman G.
'The multi-national corporation: the transfer of technology and dependence'. Development and Change, Vol.6, No.1. p.5-21. January 1975.

4.445 COLE, Sam
North South income distribution and terms of trade in a model of innovation and transfer of technology. Brighton: SPRU, 1979. 9p. Mimeo.

4.446 CONNELL, David
The UK's performance in export markets - some evidence from international trade data. London: NEDO, 1979. 54p. (Discussion paper 6).

4.447 COOPER, Charles M.
'International transactions in innovative machinery'. Journal of Development Studies, Vol.16, No.3, April 1980, p.332-351.

4.448 COOPER, Charles M. and MAXWELL, Philip
Machinery suppliers and the transfer of technology to Latin America. Brighton, SPRU, 1975. 80 leaves. Mimeo. A report by the Science Policy Research Unit to the Organisation of American States.

4.449 COOPER, Charles M. and SERCOVITCH F.
The mechanisms for the transfer of technology from advanced to developing countries. Brighton: SPRU, November 1970. 61 leaves.

4.450 COORAY, Noreen
The technological factor and its relevance to the competition between synthetic and natural rubber in international trade. Thesis presented in fulfilment of the requirements for the degree of Doctor of Philosophy, University of Sussex, August 1979. Brighton: SPRU, University of Sussex, 1979. 2 vols. Vol. 1 - 254 leaves. Vol. 2 - 550 & appendices.

4.451 COURTENAY, W.H. and LEIPZIGER, D.M.
'Multinational corporations in L.D.C.'s: the choice of technology'. Oxford Bulletin of Economics and Statistics, November 1975, p. 97-304.

4.452 CROPP, J.A.D. and others
Trade in innovation: the ins and outs of licensing. London, Wiley, 1970.

4.453 DAVIDSON, William H.
'Patterns of factor-saving innovation in the industrialized world'. European Economic Review, Vol.8, 1976, p.207-217.

4.454 DOSI, Giovanni
Trade, international adjustment, technical innovations. Brighton: Sussex Eoropean Research Centre, University of Sussex, 1981. Mimeo.

4.455 DOSI, Giovanni
International specialisation, technical change and industrial interdependences: some notes on economic signals and context conditions. Brighton: SPRU, 1982. Mimeo.

4.456 DRISCOLL, Robert E. and WALLENDER, Harvey W. (eds)
Technology transfer and development: an historical and geographic perspective. New York: Fund for Multinational Management, 1974. 301p.

4.457 ELLIOTT, Robert F. and WOOD W. Peter
'The international transfer of technology and Western European integration'. Research in International Business and Finance, Vol.12, 1981, pp.117-150.

4.458 ELLIOTT, Robert F. and WOOD, Peter W.
The international transfer of technology and Western European integration. Aberdeen: University of Aberdeen, Department of Political Economy, 1979. 65 leaves. Mimeo.

4.459 EVENSON, Robert E.
International invention: implications for technology market analysis. New Haven: Economic Growth Center, Yale University, 1982. 35p. Mimeo. (Center Discussion Paper no. 419.)

4.460 EWING, David W. (ed)
Science policy and business: the changing relation of Europe and the United States. The John Diebold Lectures, 1971. Boston: Harvard U.P., 1973. 110p.

4.461 FAUST, R.E.

'Acquisition/licensing – strategies for pharmaceutical products'. Drug and Cosmetic Industry, Vol.113, No. 4, October 1973. p.48–52.

4.462 FIELD, A.
Analysis of the changing pattern of international trade in machines and equipment, 1963–1976. Geneva: ILO, 1979. 181p. Mimeo. (International Division of Labour Programme) (World Employment Programme Research Working Papers) (WEP 2–36/WP 6).

4.463 FIELD, A.
An analysis of the changing patterns of international trade in textile and clothing. Geneva: ILO, 1979. 191p. Mimeo. (International Division of Labour Programme) (World Employment Programme Research Working Papers) (WEP 2–36/WP 3).

4.464 FINAN, William F.
The international transfer of semiconductor technology through U.S.-based firms. New York: National Bureau of Economic Research, 1975. 147p. (NBER working paper no. 118).

4.465 FREEMAN, C.
'Technical innovation and British trade performance'. De-industrialization, edited by Frank Blackaby. London: Heinemann Educational Books, 1979, pp.56–77.

4.466 FREEMAN, C. and others
'Chemical process plant: innovation and the world market'. National Institute Economic Review, No.45, August 1968, pp. 29–57.

4.467 FRONKO, Edward G.
'Licensing technology'. Industrial Research, Vol. 11(8), August 1969, pp.54–56.

4.468 GOSH, S.K.
'Transfer of military technology to China'. IDSA Journal, Vol.9, No.3, 1977, p. 220–36.

4.469 GRANDSTRAND, Ove
The role of technology trade in Swedish companies. Paper presented at a conference on Technology Transfer at the International Institute of Management Science Center, Berlin, December 8–10, 1980. Goteborg, Sweden: Chalmers University of Technology, Department of Industrial Management, 1981. 45p. Mimeo.

4.470 GREEN, John C.

'Emerging restrictions on transfer of technology'. Idea, 1971, Vol. 15, No. 2, p. 274-280.

4.471 GRUBER, William H., MEHTA, Dileep and VERNON, Raymond
'The R & D factor in international trade and international investment of United States industries'. Journal of Political Economy, February 1967, p. 20-37.

4.472 HARMAN, Alvin J.
The international computer company: innovation and comparative advantage. Cambridge, Mass.: Harvard University Press, 1971. 181p.

4.473 HAWKINS, Robert G. and GLADWIN, Thomas N.
Conflicts in the international transfer of technology: a U.S. home-country view. Paper prepared for the International Conference on Technology Transfer Control System: issues, perspectives, implications, Philadelphia, 9-10 February, 1979. 108p. Mimeo.

4.474 HAYDEN, Eric W. and NAU, Henry R.
'East-West technology transfer: theoretical models and practical experiences'. Columbia Journal of World Business, Fall 1975. Not paginated.

4.475 HIRSCH, Seev
'The United States electronics industry in international trade'. National Institute Economic Review, No. 34, 1965, 92-97.

4.476 HIRSCH, Seev
'Capital or technology? Confronting the neo-factor proportions and neo-technology accounts of international trade.' Weltwirtschaftliches Archiv, Band 110, Heft 4, 1974. Seiten 535-536.

4.477 HIRSCH, Seev
The product cycle model of international trade: a multi-country cross-section analysis. Revised February 1975. 28 leaves. Mimeo.

4.478 HIRSCHEY, Robert C. and CAVES, Richard E.
'Research and transfer of technology by multinational enterprises'. Oxford Bulletin of Economics and Statistics, Vol. 43, No. 2, May 1981, p. 115-130.

4.479 HORN, Ernst-Jurgen
Technology and international competitiveness: the German evidence and some overall comments. Paris: OECD, 1981. 42p.

Mimeo. (DSTI/SPR/81.60).

4.480 KOHLER, Barbara M., RUBENSTEIN, Albert H. and DOUBS, Charles E.
A behavioural study of international technology transfer between the United States and West Germany. Evanston, Ill.: Northwestern University, 1972. 31 leaves. (Program of research on the management of Research & Development - Project 8 (2)). Mimeo.

4.481 KRAUSE, Lawrence B.
European economic integration and the United States. Washington: Brookings Institute, 1968. 265p.

4.482 KRUGMAN, Paul
'A model of innovation, technology transfer and the world distribution of income. Journal of Political Economy, Vol. 87, No. 2, April 1979, p. 253-266.

4.483 LAKE, Arthur
Transnational activity and market entry in the semi-conductor industry. New York: National Bureau of Economic Research, Inc., 1976. 77p. (Working paper no. 126)

4.484 LIGHTMAN, Joseph M.
'Compensation patterns in U.S. foreign licensing'. Idea, Vol. 14, No. 1, 1970, p. 1-33.

4.485 LIPSEY, Robert E.
Impact of multinational firms on technology and trade flows. New York: National Bureau of Economic Research, Inc., 1976. 22 leaves. Mimeo.

4.486 LIPSEY, Robert E. and WEISS, Merle Yahr
Multinational firms and the factor intensity of trade. Washington D.C.: National Bureau of Economic Research, Inc., 1973. 28 leaves. (Working Paper no. 8) Mimeo.

4.487 LIPSEY, Robert E. and WEISS, Merle Yahr
Exports and foreign investment in manufacturing industries. New York: National Bureau of Economic Research, Inc., 1976. 66p. (Working paper no. 131)

4.488 LOVELL, Enid Baird
Foreign licensing agreements. I. Evaluation and planning. New York, National Industrial Conference Board, 1958. Studies in Business Policy, 86.

4.489 LOVELL, Enid Baird
Foreign licensing agreements. II. Contract negotiation and

administration. New York, National Industrial Conference Board, 1959. Studies in Business Policy, 91.

4.490 LUTZ, Chistian
'The 'technological gap' analysis of a European nightmare'. Swiss Review of the World Affairs, February 1968, Vol. 17(11), 9-11.

4.491 MAHONEY, James E.
Promising approaches toward understanding technology transfer. Washington, George Washington University, 1967. Program of Policy Studies in Science and Technology, Staff Discussion Paper, 201.

4.492 MAJUMDAR, Badiul Alan
Innovations, product developments and technology transfers: an empirical study of dynamic competitive advantage, the case of electronic calculators. Washington, D.C.: The University Press of America, 1982.

4.493 MANSFIELD, Edwin
Studies of the relationship between international tech-nology transfer and R & D expenditures by U.S. firms. Phila-delphia: University of Pennsylvania. No date. 103 leaves. Mimeo.

4.494 MANSFIELD, Edwin
'Economic impact of international technology transfer'. Research Management, Vol. 17, No. 1, January 1974, p. 7-11.

4.495 MANSFIELD, Edwin, ROMEO, Anthony and WAGNER, Samuel
'Foreign trade and U.S. research and development'. The Review of Economics and Statistics, Vol. LXI, No. 1, February 1979, p. 49-57.

4.496 MOMIGLIANO, Franco
Technological innovation, international trade and direct foreign investment: old and new problems for economic theory and empirical research. Paris: OECD, 1981. 47p. Mimeo. (DSTI/SPR/81.59)

4.497 McCULLOCH, Rachel
Research and development as a determinant of U.S. international competitiveness. Washington, D.C.: National Planning Association, 1978. 49p.

4.498 NATIONAL ACADEMY OF ENGINEERING.
Technology and international trade Proceedings of a symposium. October 14 & 15, 1970.Washington: National Academy of Engineering, 1971. 146p.

4.499 NATIONAL ACADEMY OF SCIENCES
The international technology transfer process. By ROBOCK,
Stephan H. and CALKINS, Robert D. Washington, D.C.:
National Academy of Sciences, 1980. 34p.

4.500 NAU, Henry R.
Technology transfer and U.S. foreign policy.   New   York,
London: Praeger, 1976. 325p.

4.501 NELSON, Richard R. and NORMAN, Victor D.
Technical change and factor mix over the product cycle: a
model of dynamic comparative advantage. Bergen: Norwegian
School of Economics and Business Admin., Institute of
Economics, 1973. 40 leaves. Norwegian School of Economics
and Business Admin., Institute of Economics, discussion
papers, 05/73)

4.502 NEU, Werner
A model of international trade & production with product
differentiation. Paper presented at the Sixth Conference of
the European Association for Research in Industrial
Economics, Paris, 9-11 September, 1979. University of Bonn,
1979. 21 leaves. Mimeo.

4.503 PAVITT, Keith
'Technology, international competition and economic growth:
some lessons and perspectives'. World Politics, 25(2),
January 1973, pp.183-205.

4.504 PEARSON Ruth
Technology, innovation and transfer of technology in the
cement industry. Buenos Aires: IDB/ECLA Research Programme
in Science and Technology, 1976. 50p. Mimeo.

4.505 PIEKARZ, Rolf
Assessing effects of international technology transfer on
the U.S. economy: public policy context, information needs
and research program. Preliminary, June, 1975. 28 leaves.

4.506 POSNER, M.
'International trade and technical change'. Oxford
Economic Papers, 1961, 13(3), 323-341.

4.507 POTTER, F.E.
U.S. licensing relations with the U.K. and Common Market.
Mimeo. Paper presented at the Management of International
Licensing Conference, 14 June, 1968.

4.508 QUINN, James Brian
'Technology transfer by multinational companies'. Harvard

Business Review, November/December 1969, 147-161.

4.509 ROSENBERG, Nathan
The international transfer of industrial technology: past
and present. Stanford, California: Stanford University,
1978. 54p. Mimeo.

4.510 ROTHWELL, Roy
'Non-price factors in the export competitiveness of
agricultural engineering products'. Research Policy, Vol.10,
No.3, July 1981, pp.261-288.

4.511 SCHNORRENBERG, Gunther
Price competitiveness in international trade: the case of
the textile and clothing industry in Denmark, Norway and
Sweden. Paper prepared for the sixth Conference of the
European Association for Research in Industrial Economics in
Cergy/France, September, 9-12, 1979. University of Kiel,
1979. 32 leaves. Mimeo.

4.512 SCIBERRAS, Ed.
International competitiveness and technical change: a study
of the US consumer electronics industry. A study prepared
for the US Department of Commerce Economic Development
Administration under agreement to Northwestern University.
Brighton: SPRU, September 1979. 59 leaves. Mimeo.

4.513 SENKER, Peter J.
'Technical change, employment and international
competititon'. Futures, Vol.13, No.3, June 1981, pp.159-170.

4.514 SOETE, Luc L.G.
Inventive activity, industrial organisation and
international trade - presented in partial fulfilment of
requirements for the degree of Doctor of Philosophy in
Economics, University of Sussex, June 1978. 375 leaves.

4.515 SOETE, Luc L.G.
The impact of technological innovation on international
trade patterns: the evidence reconsidered. Paper presented
at the Science and Technology Indicators Conference, 15-19
September, 1980, OECD, Paris. Paris: OECD, 1980. 74p.
Mimeo.

4.516 SOETE, Luc L.G.
'A general test of technological gap trade theory'.
Weltwirtschaftliches Archiv, Band 117, Heft 4, 1981,
pp.638-660.

4.517 STEWART, Frances

International technology transfer: issues and policy options. Washington D.C.: The World Bank, 1979. 118p. Mimeo. (World Bank Staff Working Paper No. 344).

4.518 SURREY, A.J. and CHESSHIRE, J.H.
The world market for electrical power equipment: rationalisation and technical change. Brighton: SPRU, 1972.

4.519 TEECE, David J.
The multinational corporation and the resource cost of international technology transfer. Cambridge, Mass.: Ballinger, 1977. 129p.

4.520 TEUBAL, Morris
Primary exports and economic development: the role of the engineering sector. Jerusalem: The Maurice Falk Institute for Economic Research in Israel, 1979. 23p. Mimeo. (Discussion Paper No.7915).

4.521 TIMBERG, Sigmund
The role of the international patent system in the international transfer and control of technology. Paper presented at the International Conference on Technology Transfer Control Systems: issues, perspectives, implications, Philadelphia, 9–10 February 1979. No imprint. 1979. 34p. Mimeo.

4.522 UNITED STATES – Department of Labor
Summaries of recent analyses on U.S. international competitiveness and the changing structure of U.S. trade. Submitted to the Subcommittee on Industrial Growth and Productivity Senate Committee on the Budget, December 3, 1980. Washington D.C.: U.S. Department of Labor, 1980. 21 leaves. Mimeo.

4.523 UNITED STATES – National Academy of Sciences
Technology, trade and the U.S. economy. Report of a workshop held at Woods Hole, Massachusetts, August 22–31 1976. Washington D.C.: National Academy of Sciences 1978. 169p.

4.524 VERNON, Raymond (ed)
The technology factor in international trade. A conference of the universities – National Bureau Committee for Economic Research. New York: N.B.E.R. 1970. (Distributed by Columbia U.P.). 493p.

4.525 VON BERTRAB-ERIMAN, Herman Raimund
The transfer of technology: a case study of European private enterprises having operations in Latin America with special emphasis on Mexico. A dissertation. University of

Texas, 1968. 260 leaves. Mimeo.

4.526 WALKER, W.B.
Industrial innovation and international trading performance.
Greenwich, Conn.: JAI Press 1979. 122p.

4.527 WILLS, Russel M.
The international transfer and licensing of technology in
Canada. Draft. Ottawa: Department of Industry, Trade and
Commerce, 1982. 126p. Mimeo.

4.d Technical Change & Regional Economic Development

4.528 CLARK, Norman G.
'Science, technology and regional economic development'.
Research Policy, Vol.1, No.3, July 1972, p.296-319.

4.529 COOMBS, R. and GREEN, K.
Regional analysis of the impact of technical change on
employment. Manchester: UMIST, Department of Management
Sciences, 1982. 16 leaves. Mimeo. (Occasional Paper
No.8204).

4.530 EWERS, H.J. and WETTMANN, R.W.
'Innovation-oriented regional policy'. Regional Studies,
Vol.14, No.3, p. 161-179.

4.531 GIBBS, D.C. and EDWARDS, A.
The interregional diffusion of selected process innov-
ations: some preliminary evidence. Paper given to Regional
Science Association, Durham, 3 September 1981. New-
castle-upon-Tyne: Centre for Urban and Regional Development
Studies, University of Newcastle-upon-Tyne, 1981. 18 leaves.
Mimeo.

4.532 GIBBS, D.C. and EDWARDS, A.
Regional variations in the rate of introduction of new
technology in manufacturing industry. Paper presented in
Durham University Business School Research Seminar Series,
18 February 1982. Newcastle: Centre for Urban and Regional
Development Studies, University of Newcastle-upon-Tyne,
1982. 27 leaves. Mimeo.

4.533 GIBBS, D.C., EDWARDS A. and THWAITES, A.T.
Factors affecting the inter-regional diffusion of new
production technology: the UK experience. Paper presented at
the Institute of British Geographers' Annual Conference,

held at Southampton University, 6 January 1982. Newcastle-upon-Tyne: Centre for Urban and Regional Development Studies, University of Newcastle-upon-Tyne, 1982. 19 leaves. Mimeo.

4.534 GODDARD, J.B.
Industrial innovation and regional economic development in Britain. Paper presented to the Commission in Industrial Systems, International Geographical Union Congress, Tokyo, August, 1980. Newcastle Upon Tyne: University of Newcastle Upon Tyne, Centre for Urban and Regional Development Studies, 1980. 31p. & Tables. Mimeo.

4.535 GODDARD, J.B. and THWAITES, A.T.
Technological change and the inner city. SSRC Inner Cities in Context contractor's report. University of Newcastle Upon Tyne, Centre for Urban & Regional Development Studies, 1979. 81p.

4.536 GREEN, Kenneth, COOMBS, Rod and HOLROYD, Keith
The effects of microelectronic technologies on employment prospects: a case study of Tameside. A report commissioned by Tameside Metropolitan Borough Council. Farnborough, Hants: Gower, 1980. 133p and Appendices.

4.537 MALECKI, Edward J.
Science, technology and regional development: a review and prospects. Norman, Oklahoma: Department of Geography, University of Oklahoma. No date. 38 leaves. Mimeo.

4.538 NORTH EAST LONDON EMPLOYMENT GROUP
Microelectronics and the North-East London economy - impact and action: a NELEG discussion document. London: NELEG, 1980. 34 leaves. Mimeo.

4.539 OAKEY, R.P.
An analysis of the spatial distribution of significant British industrial innovations. Newcastle: Centre for Urban and Regional Development Studies, University of Newcastle, 1979. 42 leaves. Mimeo. (Centre for Urban and Regional Development Studies, Discussion Paper No. 25).

4.540 OAKEY, R.P.
'The regional distribution of innovative manufacturing establishments in Britain'. Regional Studies, 1980, Vol. 14, pp.235-253.

4.541 ROTHWELL, Roy
Innovation, small firms and regional development. Paper prepared for the Six Countries Programme on Aspects of

Government Policy towards Innovation, Workshop on Regional Innovation Policy, Sophia Antipolis, 1st and 2nd June, 1981. Brighton: SPRU, February 1981. 16p. Mimeo.

4.542 ROTHWELL, Roy
'The role of technology in industrial change: implications for regional policy', Regional Studies,Vol.16, No.5, 1982, pp.361-369.

4.543 THWAITES, A.T.
Technological change, mobile plants and regional development. Newcastle: Centre for Urban and Regional Development Studies, University of Newcastle, 1977. 32p. Mimeo.

4.544 THWAITES, A.T.
The future development of R & D activity in the Northern region: a comment. Newcastle: Centre for Urban and Regional Development Studies, 1978. 19 leaves. Mimeo.

4.545 THWAITES, A.T., OAKEY, Raymond and NASH, Peter
Industrial innovation and regional development. Vol.1. Final report to the Department of the Environment. Newcastle-Upon-Tyne: University of Newcastle-Upon-Tyne, Centre for Urban and Regional Development Studies, 1981. 98 leaves. Mimeo.

CHAPTER 5. <u>Social And Political Aspects Of Technical Change</u>

5.a The Decision-making Environment Of Innovation

5.001 ALBURY, David
<u>Partial progress: the politics of science and technology.</u>
London: Pluto Press,1982. 215p.

5.002 AUSTRALIAN POSTAL AND TELECOMMUNICATIONS UNION
<u>Submission by the Australian Postal and Telecommunications</u>
<u>Union to the Committee of Inquiry into Technological Change</u>
<u>in Australia.</u> Carlton South, Victoria: APTU, 1979. 50
leaves. Mimeo.

5.003 AYERS, Eugene
'Social attitudes toward invention'. <u>Science in Progress</u>,
1957, 10th series, 47-73.

5.004 BRAENDGAARD, Asger
<u>The welfare state and social technology: some options in</u>
<u>the application of microelectronics and the associated</u>
<u>benefits and risks.</u> Paper prepared for a EEC/FAST
conference, London, 25-29 January 1982. Aalborg: Aalborg
University Center, Institute of Production, 1981. 18p.
(Smaskrift no 29).

5.005 BRIGHT, Willard M.
'Innovators - born or made?' <u>Technology and Society</u>, 1969,
4(4), 29-31.

5.006 BROOKS, Harvey
'Technology, evolution and purpose'. <u>Daedalus</u>, Vol.109,
No.1, Winter 1980, p.65-81.

5.007 CANADA - Department of Industry, Trade and Commerce:
       Technology Branch
<u>The attitude of trade unions towards technological changes.</u>
By S.G. Peitchinis. Ottawa: Department of Industry, Trade
and Commerce, 1980. 73p. (Technological Innovation Studies
Program, Research Report.)

5.008 CANADA - Department of Labour Economics and Research

The collective agreement in Canada: the study of its role in a changing industrial environment. A series of eight papers by F. Quinet. Ottawa, 1967.

5.009 CANADA - Science Council of Canada
Planning now for an information society: tomorrow is too late. Ottawa: Science Council of Canada, 1982. 77p. (Science Council of Canada Report 33).

5.010 CONSUMER INTERESTS AND MANUFACTURING CONSIDERATIONS
A dialogue between industry, government and society. 11th International T.N.C. Conference Rotterdam, 23-24 Feb. 1978. The Hague: Netherlands Central Organization for Applied Scientific Research TNC, 1978. 94p.

5.011 COOPER, Charles M.
A case study of choices of techniques in two processes in the manufacturing of cans. SPRU, September 1972. Mimeo. (Technical paper 7).

5.012 COOPER, Charles M.
Choice of techniques and technological change as problems in political economy. March 1973. 15 leaves. Mimeo.

5.013 CURNOW, R, KALDOR, M, McLEAN, M, ROBINSON, J and
        SHEPHERD, P
'General and complete disarmament: a systems analysis approach', Futures, Vol.8,No.5, October 1976, pp.384-391.

5.014 DAVIS, Vincent
The politics of technological innovation: patterns in navy cases. Paper presented to the AIAA third annual meeting, Boston, Mass., 29 November - 2 December 1966. AIAA Paper no.66-984. New York, American Institute of Aeronautics and Astronautics, 1966.

5.015 DICKSON, David
Alternative technology and the politics of technical change. London: Fontana/Collins, 1974. 224p.

5.016 DIEBOLD, John
'Is the gap technological?' Foreign Affairs, January 1968, Vol.46(2), 276-291.

5.017 ESIST SEMINAR
European Society and its Interactions with Science and Technology. Compiegne, France, 19-20 October 1978. Various papers held.

5.018 GREAT BRITAIN - Department of Industry

Social aspects ot innovation: a survey of research. By Lewis
F. Abbott. London: HMSO, 1976. 131p. (Dept. of Industry
CFIT Paper no. 1).

5.019 KALDOR, Mary and PERRY-ROBINSON, Julian
'War',in Freeman, C. and Jahoda, M.(eds.) World futures:
the great debate, Oxford: Martin Robertson, 1978.

5.020 KROHN, Wolfgang, LAYTON, Edwin T. and WEINGART,
        Peter (eds)
The dynamics of science and technology: social values,
technical norms and scientific criteria in the development
of knowledge. Dordrecht, Holland: Reidel Publishing
Company, 1978. 293p.

5.021 MITCHELL, Jeremy.
'The consumer movement and technical change', International
Social Science Journal, 1973, no. 3, pp. 358-369.

5.022 MORTIMER, J. E.
Trade unions and technological change. London: O.U.P.,
1971. 108p.

5.023 MULKAY, M. J.
The social process of innovation: a study in the sociology
of science. London: MacMillan, 1972. 64p.

5.024 McLAUGHLIN, Doris B.
The impact of labor unions on the rate and direction of
technological innovation. Ann Arbor, Michigan: Institute of
Labor and Industrial Relations, The University of Michigan -
Wayne State University, 1979. 148p. Mimeo.

5.025 PERRY-ROBINSON, Julian
'Chemical arms control and the assimilation of chemical
weapons', International Journal, Vol. 36, No. 3, Summer
1981.

5.026 RAHMAN, A.
'Science, technology and development: a socio-historical
analysis and possible future lines of action'. The Journal
of Scientific and Industrial Research, Vol.37, No.12, 1978,
pp.635-644.

5.027 RAMO, Simon
America's technology slip. New York: John Wiley, 1980.
29/p.

5.028 RAVETZ, Jerome
'Scientific knowledge and expert advice in debates about

large technological innovations'. _Minerva_, Vol.XVI, No.2, Summer 1978, pp.273-282.

5.029 ROTHSCHILD, Joan(ed.)
Women, technology and innovation. London: Pergamon, 1982.

5.030 SINCLAIR, Craig, MARSTRAND, Pauline and NEWICK, Pamela
Innovation and human risk: the evaluation of human life and safety in relation to technical change. London: Centre for the Study of Industrial Innovation, 1972. 36p.

5.031 WALKER, W.B. with LOENNROTH, M.
Nuclear power struggles: industrial competition and proliferation control. London: Allen and Unwin,1983.

5.b Ecological Environment, Technical Change & Innovation

5.032 CONFEDERATION OF BRITISH INDUSTRY
Technology and the environment. Three papers specially prepared for discussion at a Conference held 22 September 1970. London: C.B.I., 1970. 45p.

5.033 EICHHLOZ, Geoffrey G.
Environmental aspects of nuclear power. Ann Arbor, 1976. 683p.

5.034 EIRMA
Industrial R & D and environmental protection. Conference Papers, EIRMA Annual Conference, Copenhagen, 1973. Paris: EIRMA, 1973. 115p. (Conference papers series vol. XIV).

5.035 GREAT BRITAIN
Nuclear power and the environment. The government's response to the sixth report of the Royal Commission on environmental pollution (Cmnd. 6618). London: HMSO, 1977. 21p. Cmnd. 6820.

5.036 GREAT BRITAIN - Department of Industry
Technology and the environment: noise. Part 1. Denmark, Finland, Norway and Sweden. London: Department of Industry, 1979. 52p. Pbk. (Reports from overseas posts: No.11).

5.037 GREAT BRITAIN - Royal Commission on Environmental Pollution
Sixth report: nuclear power and the environment. Chairman, Sir Brian Flowers. London: HMSO, 1976. 237p. Command 6618.

5.038 HILL, Christopher, T.
<u>A state ot the art review of the effects of regulation on</u>
<u>technological innovation in the chemical and allied products</u>
<u>industries.</u> CAPI Project, St. Louis: Center for Development
and Technology, Washington University, 1975. (Vol. 1:
Executive summary. Vol. 2 The State of the Art. Vol. 3
Abstracts and literature list).

5.039 JURGENSEN, Harald
<u>Effects ot measures for protection of the environment on</u>
<u>industrial development and the siting of undertakings.</u> Paper
presented at the Conference on Industry and Society in the
European Community, Venice, 1972. 45p. (Report no. 6.)

5.040 LAOT, J.
<u>Implications of environmental measures for industrial</u>
<u>development and the siting of enterprises.</u> Paper presented
at the Conference on Industry and Society in the European
Community, Venice, 1972. 50p. (Report no. 6.)

5.041 LEDERMAN, Leonard L. and MORRISON, Richard E.
<u>Environmental and worker health/safety regulation and</u>
<u>technological innovation.</u> Paper prepared for the
International Conference on Regulation and Innovation, The
Hague, June 11-13, 1979. Washington, D.C.: N.S.F., 1979. 26
leaves. Mimeo.

5.042 McROBIE, George
<u>Small is possible.</u> London: Jonathan Cape, 1981. 331p.

5.043 PELTZMAN, Sam
<u>Regulation of pharmaceutical innovation: the 1962</u>
<u>amendments.</u> Washington, D.C.: American Enterprise Institute
for Public Policy Research, 1974. 118p. Pbk.

5.044 SCHWEITZER, Glenn E.
<u>Regulation and innovation: the case of environmental</u>
<u>chemicals.</u> Cornell University, Program on Science,
Technology and Society, 1978. 19 leaves. Mimeo.

5.045 SIX COUNTRIES PROGRAMME
<u>Innovation in industry in relation to regulations in the</u>
<u>areas of environmental protection, occupational health and</u>
<u>safety and in telecommunications.</u> Papers presented at the
joint workshop of the Six Countries Programme and N.S.F.,
Washington, D.C., held in The Hague, June 11-13, 1979.
Delft: Secretariat of the Six Countries Programme, 1979.
183p.

5.046 UNITED NATIONS - Economic Commission for Europe

Environment and energy: environmental aspects of energy production and use with particular reference to new technologies. Oxford: Pergamon   Press   for   the   United Nations, 1979. 113p.

5.c Social Effects Of Technical Change & Innovation

5.04/   BATTELLE COLUMBUS LABORATORIES
Science, technology and innovation,   prepared   for   the National   Science   Foundation. Columbus,   Ohio: Battelle Columbus Laboratories, 1973. 33p.

5.048   BEANLAND, D.G. et al.
Technological change - impact of information technology.
Canberra: Information Technology Council, 1980. 138 leaves. Mimeo. (PBB1-125 494).

5.049   BERTING, J., MILLS, S.C. and WINTERSBERGER, H. (eds)
The socio-economic impact of microelectronics.   Oxford: Pergamon Press, 1980. 267p.

5.050   BJORN-ANDERSEN, Niels, EARL, Michael, HOLST, Olav
        and MUMFORD, Enid (eds.)
Information Society: for richer, for poorer. Selected papers from a conference, Selsdon Park Hotel, London. 25-29 January 1982. Amsterdam: North Holland. 1982.

5.051   BRADY, T. and MILES, I.
'Robots and their social impact- the U.K. case'. in European Pool of Studies, Changement social et technologie en Europe, Bulletin of Information No. 10 November-December,1982.

5.052   BRITISH ASSOCIATION FOR THE ADVANCEMENT OF SCIENCE
Automation - friend or foe? One day symposium, 8 June 1978. London: B.A.A.S., 1978. 36p.

5.053   CROSSMAN, Edward R.F.W.
Taxonomy of automation: state of arts and prospects.   OECD, European   Conference   on   Manpower Aspects of Automation and Technical Change, Zurich, 1-4 February 1966.

5.054   FORESTER, Tom (ed)
The microelectronics revolution: the complete guide to the new technology and its impact on society.   Oxford: Blackwell, 1980. 589p.

5.055   FREEMAN, C.

Social and economic impact of microelectronics. Paper presented at the 1979 Workshop on Technology Assessment - its role in national and corporate planning, held Sydney 25-26 July 1979 and organised by the Department of Science and the Environment. Canberra: Australian Government Publishing Service, 1979. pp. 13-24.

5.056 FRIEDRICHS, Gunter and SCHAFF, Adam (eds)
Microelectronics and society: for better or for worse. A report to the Club of Rome. Oxford: Pergamon Press, 1982. 353p. (Pergamon International Library of Science, Technology, Engineering and Social Studies)

5.057 GENERAL TECHNOLOGY SYSTEMS LTD.
Netherlands microelectronics study. Study carried out at the request of the "Advisory Group for the Societal Consequences of Microelectronics" and administered by the Ministry of Science Policy. London: General Technology Systems Ltd., 1979. Not paginated. (GT 78035).

5.058 GORDON, Theodore J. and AMENT, Robert H.
Forecasts ot some technological and scientific developments and their societal consequences. Middletown, Conn., Institute for the Future, 1969.

5.059 GREAT BRITAIN - Central Policy Review Staff
Social and employment implications of microelectronics. London: NEDC, 1978. 24p. Mimeo.

5.060 HILLS, Philip (ed)
The future of the printed word: the impact and the implications of the new communications technology. London: Frances Pinter, 1980. 172p.

5.061 HINES, Colin, BENNETT, Peter, PELTU, Malcolm and POPAY, Jennie
Beyond generalisations: issues in the new technology debate. A collection of papers arising from a series of seminars. London: Polytechnic of the South Bank, Department of Town Planning, 1981. 85p. Mimeo. (Earth Resources Research and Polytechnic of the South Bank Planning Department Planning Conference Paper 14).

5.062 HOFFMAN, Kurt and MILES, Ian
New (information) technology and social change in the United Kingdom. Brighton: SPRU, January 1982. 29p. Mimeo.

5.063 MARCSON, Simon
Automation, alienation and anomie. New York, Harper, 1970.

5.064 MARSTRAND, Pauline
'Ecological and social evaluation of industrial development'. Environmental Conservation, 3(4), pp.303-308.

5.065 NORA, Simon and MINC, Alain
The computerization of society. A report to the President of France. Cambridge, Mass./London, MIT Press, 1980. 186p. (Originally published as L'information de la societe, 1978. Paris: Documentation Francais).

5.066 RADA, J.
The impact of micro-electronics: a tentative appraisal of information technology. Geneva: ILO, 1980. 109p. (A WEP study).

5.067 RATHENAU ADVISORY GROUP
The social impact of micro-electronics. Report of the Rathenau Advisory Group. The Hague: Government Publishing Office, 1980. 111p.

5.068 ROSENBROCK, H.H.
Robots and people. Draft. Fourth Harley Lecture to be presented 24 November 1981. (No imprint). 1981. 16p. Mimeo.

5.069 SENKER, P.
'Social implications of automation'. The Industrial Robot, June 1979, pp.59-61.

5.070 TAYLOR, David
Cheap words...? Word processing: past, present and future. London: Youthaid, 1979. 21 leaves. Mimeo.

5.071 TURN, Rein
Computers in the 1980's. New York, London: Columbia U.P., 1974. 257p.

5.072 WHISTON, Tom
'Microelectronics: some wider implications'. Chemistry and Industry, September, 1982, pp.629-635.

CHAPTER 6. Management Of Technical Change And Innovation

6.a Organisation & Management Of R&D & Scientific
    Laboratories

6.001 ALLEN, Thomas J.
The utilization of information sources during R & D
proposal preparation. Cambridge, Mass: Sloan School of
Management, 1964. Research Program on the Management of
Science and Technology, paper 97-67.

6.002 ALLEN, Thomas J.
Problem solving strategies in parallel research and
development projects. Cambridge, Mass.: Sloan School of
Management, 1965. Research Program on the Management of
Science and Technology, paper 126-65.

6.003 ALLEN, Thomas J.
Sources of ideas and their effectiveness in parallel R & D
projects. Cambridge, Mass.: Sloan School of Management,
1965. Research Program on the Management of Science and
Technology, paper 130-65.

6.004 ALLEN, Thomas J.
'Communication in the research and development laboratory'.
Technology Review, November 1967, 70(1), 31-37.

6.005 ALLEN, Thomas J.
Managing the flow of technology: technology transfer and
the dissemination of technological information within the R
& D organization. Cambridge, Massachusetts and London: The
MIT Press, 1977. 311p.

6.006 ALLEN, Thomas J. and COONEY, Sean
Institutional roles in technology transfer: a diagnosis of
the situation in one small country. Cambridge, Mass.: Sloan
School of Management, 1973. 31p. Mimeo.

6.007 ALLEN, Thomas J. and REILLY, Valentine
Getting the world around: report of a pilot study on
technology transfer to Irish industry. Cambridge, Mass.:
Sloan School of Management, 1973. 21 leaves. Mimeo.

6.008 ALLEN, Thomas J. and others
Time allocation among three technical information channels
by R & D engineers. Cambridge, Mass.: Sloan School of
Management, 1966. Research Program on the Management of
Science and Technology, paper 184-66.

6.009 ALLISON, David
'The civilian technology lag'. International Science and
Technology, December 1963, No.24, pp.24-34.

6.010 ALLISON, David (ed)
The R & D game: technical men, technical managers, and
research productivity. Cambridge, Mass.: MIT Press, 1969.

6.011 ANDERSON, Clifton P. and RAMEY, James T.
'Congress and research: experience in atomic research and
development'.   Annals of American Academy of Political
Science, 1960, 327, 85-94.

6.012 ANDREW, G.H.L.
'Assessing priorities for technical effort'. Operational
Research Quarterly, September 1954, 5(3), 67-80.

6.013 ANDREWS, Frank M. and FARRIS, George F.
'Supervisory practices and innovation in scientific teams'.
Personnel Psychology, 1967, 20(4), 497-515.

6.014 ANNERSTEDT, Jan
A survey of world research and development efforts: the
distribution of human and finanacial resources devoted to
research and experimental development in 1973.   Roskilde,
Denmark: Roskilde University, Institute of Economics and
Planning, 1979. 72p.

6.015 ANNERSTEDT, Jan
Indigenous R & D capacities and international diplomacy:
some implications of the present global distribution of R &
D resources. Roskilde: Roskilde University Center, 1979.
36p.

6.016 ARNON, I.
Organisation and administration of agricultural research.
London: Elsevier, 1968. 80p.

6.017 ARTHUR D. LITTLE, INC.
Federal funding of civilian research and development. Vol.1
- Summary. Vol.2 - Case Studies. Prepared for Experimental
Technology Incentives Program. N.B.S. Washington, D.C.:
Arthur D. Little, Inc., 1976. 55p., 353p.

6.018 ASLIB
Accelerating innovation. Papers given at a symposium held at the University of Nottingham, March 1969. London, 1970.

6.019 AUSTRALIAN ACADEMY OF SCIENCE - SCIENCE AND INDUSTRY
        FORUM
Obstacles and aids to innovation. Papers delivered at the 7th Forum meeting, 28 February 1970. Canberra, 1970. (Report No. 2).

6.020 AVOTS, Ivars
'Why does project management fail?' California Management Review, Fall 1969, 12(1), 77-82.

6.021 BARTH, Richard T.
'The configuration and quality of intergroup climates as perceived by engineers and scientists'. Studies in Personnel Psychology, October 1971, Vol. 3, No.2, p.69-81. (Program of Research on the Management of Research and Development - Project 8 (1).

6.022 BARTH, Richard T.
A comparison of weighted and unweighted intergroup climate satisfaction scores 1,2. Evanston, III: Northwestern University, 1971. 26 leaves. (Program of Research on the Management of Research and Development - Project 8 (1). Mimeo.

6.023 BARTH, Richard T.
Description and characteristics of the intergroup climate instrument for engineers and scientists. Evanston, III: Northwestern University, 1971. NP. (Program of Research on the Management of Research and Development - Project 8(1). Mimeo.

6.024 BASS, Lawrence W. and OLD, Bruce S. (eds)
Formulation of research policies: collected papers from an international symposium. Washington, AAAS, 1967. AAAS Publication 87.

6.025 BEATTIE, C.J. and READER, R.D.
Quantitative management in R & D. London: Chapman & Hall, 1971. 374p.

6.026 BEHRMAN, J.N. and FISCHER, W.A
Overseas R & D activities of transnational companies.
Cambridge Mass.: Delgeschlager, Gunn & Hain, 1980. 344p.

6.027 BINNING, K.G.H.
The analysis of research and development in the Ministry of

Technology. Didcot, Berks, Programmes Analysis Unit, 1967.
PAU M3/67.

6.028 BLACK, Guy
Financial variables associated with R & D expenditures by
industry. Washington D.C.:   Program of Policy Studies in
Science and Technology, George Washington University, 1968.
(Staff Discussion Paper, 300).

6.029 BLACK, Guy
What's happening to small business research and development?
Washington, D.C.: Program of Policy Studies in Science and
Technology, George Washington University, 1971.   (Staff
Discussion Paper 303).

6.030 BLACK, Guy
Dialogues with management on research and development.
Washington DC: Program of Policy Studies in Science and
Technology, George Washington University, 1974. 79p.

6.031 BLACK, Guy
Output orientation in R & D - a better approach? Washington
D.C.:   Program of Policy Studies in Science and Technology,
George Washington University, 1974. 20p.

6.032 BLACK, Guy
Patterns of association in research and development.
Washington, D.C.: Program of Policy Studies in Science and
Technology, George Washington University, 1974. 55p.

6.033 BLACK, Guy
Patterns of impact and response in research and development
in industry: summary of a study. Washington, D.C., Program
of Policy Studies in Science and Technology.   George
Washington University, 1974. 17p.

6.034 BLACK, Guy
The sensitivity of types of research and development to
business conditions. Washington, D.C.:   Program of Policy
Studies in Science and Technology, George Washington
University, 1974. 23p.

6.035 BLACK, Guy and FISCHER, William A.
Research, development and financial performance. Washington,
D.C.:   Program of Policy Studies in Science and Technology,
George Washington University, 1974. 31p.

6.036 BLACK, Guy.
Research, development and business conditions, 1960-71.
Washington, D.C.: Program of Policy Studies in Science and

Technology, George Washington University, 1974. 29p.

6.037 BLAKE, Stewart P.
Managing for responsive research & development. San Francisco: W H Freeman & Co, 1978. 280p.

6.038 BLUME, Stuart S.
Structures and values as barriers to innovation-orientated
research in Britain. Paper to be presented at Symposium on
Cross National Perspectives in Industrial Research:
Obstacles and Innovation in the Applied Sciences, Paris, May
1979. 29 Leaves. Mimeo.

6.039 BOGUE, J.Yule
'The organisation and economics of research in the
pharmaceutical industry'. The Pharmaceutical Journal, 13
January 1962, 188, (5124), 4th Series. V.134, 27-32. 55-56.

6.040 BRANDENBURG, R G. and LANGENBERG, F C.
'R & D project selection and control at Crucible Steel
Corporation'. Research Management, 1969, 12(2), 123-139.

6.041 BRIGHT, James Rieser
Research, development and technological innovation: an
introduction. Homewood, III, Irwin, 1964.

6.042 BRITISH INSTITUTE OF MANAGEMENT
Management of research and development: a symposium.   By
Edward Brech and others. London, 1964.

6.043 BROEHL, Wayne G.
'A less developed entrepreneur?' Columbia Journal of World
Business, March/April 1970, p. 26-34.

6.044 BROOKS, Harvey
'Applied science and technological progress'. Science, 30
June 1967, Vol.156, 1706-1712.

6.045 BROWN, Lord Wilfred
Product design: a paper prepared for NEDO.   (No   imprint),
1977. 7 leaves. Mimeo.

6.046 BROZEN, Yale
'Trends in industrial research and development'. Journal of
Business, July 1960, Vol.33(3), 204-217.

6.047 BUMSTEAD, D.C.
Organisation for industrial R & D. Manchester: Manchester
Business School, 1969. 13 leves. (Manchester Business School
Internal Memoranda no. 27).

6.048   CANADA - Science Council of Canada
Background studies in science policy: projections of R & D
manpower and expenditure. By R.W. Jackson and others.
Ottawa, Queens Printer, 1969. Special Study No.6.

6.049   CANADA - Science Council of Canada
Background to invention: a summary of views on the Canadian
patent system and on industrial research & development
activities in Canada. By Andrew H. Wilson. Ottawa: Queen's
Printer, 1970. 77p. (Special Study no. 11.)

6.050   CHORAFAS, D.N.
Managing industrial research for profits: with case studies.
London: Cassell, 1967. 314p.

6.051   CLARK, Norman G.
The techno-economic relationship between industry and the
scientific infrastructure - with special reference to the
electronics industry and Scottish economic development.
Ph.D. thesis, University of Edinburgh, 1971. 215p.

6.052   COALES, J.
'Financial provision for research and development in
industry'. Journal of Industrial Economics, July 1957,
Vol.5(3), 239-242.

6.053   COATES, David R.
Technological forecasting and the planning of research and
development. Chilton, Berks., Programmes Analysis Unit,
1967. P.A.U.M 4/67.

6.054   COOPER, Arnold C.
'Small companies can pioneer new products'. 1966. In:
Harvard Business Review, R & D management series... Vol.2.

6.055   COOPER, Charles M.
On risky innovations and returns to R & D. Brighton: SPRU &
IDS, May 1979. 29 leaves and appendices. Mimeo.

6.056   CROSLAND, Anthony
'The falling growth of research'. Science Journal, February
1967, Vol.3(2), 80-84.

6.057   CURNOW, Ray C.
'Economies of scale in R & D'. New Scientist, 22 May 1969,
Vol.42(650), 422-423.

6.058   DAVIES, O.L.
'Some statistical considerations in the selection of projects for research in the pharmaceutical industry'.

Industrielle Organisation, 1964, Vol.33(3), 137-143.

6.059 DEAN, B.V. and GOLDHAR, J.L. (eds)
Management of research and innovation. Oxford: North-
Holland, 1980. 300p. (Studies in the Management Sciences,
Vol. 15).

6.060 DEAN, Burton V. (ed)
Operations research in research and development. Proceedings
of a conference at Case Institute of Technology. New York:
Wiley, 1963.

6.061 DEAN, Burton V. and SENGUPTA, S.S.
'Research budgeting and project selection', IRE
Transactions on Engineering Management, December 1962,
Vol.EM-9, 158-169.

6.062 DEDIJER, Stevan
'Scientific research and development: a comparative study'.
Nature, 1960, Vol. 187, 458-461.

6.063 DEDIJER, Stevan
'Research policy - its making and measurements'. Tek.
Vetenskaplig Forskning, 1963, Vol. 34(4), 134-146.

6.064 DISMAN, S.
'Selecting R & D projects for profit', Chemical Engineering,
24 December 1962, Vol.69, 87-90.

6.065 DOBROV, G.M.
A strategy for organized technology. Laxenburg: IIASA,
1977. 25 leaves. Mimeo.

6.066 DOBROV, G.M.
'The strategy for organized technology in the light of hard,
soft, and org ware interaction', Long Range Planning,
Vol.12, August 1979, pp. 79-90.

6.067 DOUDS, Charles F.
'The state of the art in the study of technology transfer -
a brief survey', R & D Management, 1971, Vol.1, No.3,
p.125-131 (Program of Research on the Management of Research
& Development - Project 8 (1)).

6.068 DROR, Israel and RUBENSTEIN, Albert H.
Top management roles in R & D projects. Evanston, Illinois:
Department of Industrial Engineering and Management
Sciences, The Technological Institute, Northwestern
University, 1981. 14 leaves & Appendices. Mimeo. (Program of
Research on the Management of Research and Development).

6.069 DUKES, Roland E., DYCKMAN, Thomas R. and ELLIOTT, John A.
'Accounting for research and development costs: the impact on research and development expenditures'. Journal of Accounting Research, Vol.18, Supplement, 1980, p. 1-37.

6.070 EIRMA
European industrial research faced with the energy crisis: problems and opportunities. Conference Papers, EIRMA Annual Conference, Paris, 13-14 March 1974. Paris: EIRMA, 1974. 103p. (Conference papers vol. 15).

6.071 EIRMA
Research and development for industry of the future. Conference Papers, EIRMA Annual Conference, Brussels, 29-31 May 1974. Paris: EIRMA, 1974. 91p. (Conference papers series vol.XVI).

6.072 EIRMA
Industrial R & D in the innovation process. Conference Papers, EIRMA Annual Conference, Stratford-Upon-Avon, 6-8 May 1981. Paris: EIRMA, 1981. 157p. (Conference Papers Vo. XXV).

6.073 ELIASSON, Gunnar and YSANDER, Bengt-Christer
Picking winners or bailing out losers? Stockholm: IUI, 1981. 70 leaves. Mimeo.

6.074 EUROPEAN COMMUNITIES - Economic and Social Committee
Organisation and management of community research and development. Brussels: General Secretariat of the Economic and Social Committee, 1980. 159p.

6.075 FABRICANT, Solomon and others
Accounting by business firms for investment in research and development. Summary, New York University, 1975. 9 leaves. Mimeo.

6.076 FALK, Charles E.
'An operational, policy-orientated research categorization scheme'. Research Policy, Vol.2, No.3, October 1973, p. 186-202.

6.077 FALK, Charles E.
'Dynamics and forecasts of R & D funding'. Technological Forecasting and Social Change, Vol. 6, No.2, 1974, p. 171-189.

6.078 FARRAR, D.J.
Product competitiveness: an opportunity for improving the

relationship between company management and engineering designers. No imprint. 19 leaves. Mimeo.

6.079 FAUST, R.E.
'Research planning: perspectives and challenges. Drug and Cosmetic Industry, Vol. 111, No. 1, July 1972. p.p.42-48. 113-114, 117-119.

6.080 FISHER, Franklin M. and TEMIN, Peter
'Returns to scale in research and development: what does the Schumpeterian hypothesis imply?' Journal of Political Economy, January/February 1973. p. 56-70.

6.081 FISHLOCK, David
'The atomic spin-off'. Management Today, March 1973, p. 51-6.

6.082 FOSTER, Richard N.
'Organize for technology transfer'. Harvard Business Review, November-December 1971. p.p.110-120.

6.083 FRANKEN, Peter
'Research   inhibitions'.   International Science and Technology, No. 17, 1963, pp.46-49.

6.084 FREEMAN, C.
'Research and development: a comparison between British and American industry'. National Institute Economic Review, No. 20, May 1962, pp.21-39.

6.085 FREEMAN, C.
'Research and development in electronic capital goods'. National Institute Economic Review, No. 34, November 1965, pp.40-97.

6.086 FREEMAN, Chris and GOLDSMITH, Maurice
'Coupling systems in industrial R & D'. In: Technological innovation and the economy. Edited by Maurice Goldsmith. London: Wiley, 1970. Chapter 18, p.177-187.

6.087 FREEMAN, Raoul J.
'A generalized network approach to project activity sequencing'. IRE Transactions on Engineering Managment, Vol. EM-7(3), 1960, pp.103-107.

6.088 FREEMAN, Raoul J.
'A stochastic model for determining the size and allocation of the research budget'. IRE Transactions on Engineering Management, Vol. EM-7(1), 1960, pp.2-7.

6.089 GABOR, Dennis
The proper priorities of science and technology.
Southampton: University of Southampton, 1972. 16p. (18th
Fawley Foundation Lecture).

6.090 GEE, Sherman
'The role of technology transfer in innovation'. Research
Management, November 1974, p.31-36.

6.091 GERMANY - Bundesminister der Verteidigung
Research and development goals of the German army.   Bonn,
1972. Microfiche. (AD 750 303).

6.092 GERRISTEN, J C.
The problems and methods of financing scientific and
technical research. Working document for meeting of experts
on the role of science and technology in economic
development, UNESCO, Paris, 11-18 December, 1968. 47 leaves.
Mimeo.

6.093 GIBB, Allan and WEBB, Terry (eds)
Policy issues in small business research.   Farnborough,
Hants: Saxon House, 1980. 214p.

6.094 GISCARD, Pierre Henri
Optimal parameters of the organisation of research and of a
research career system from the standpoint of the efficiency
of research and of research workers. Paris, Unesco, 1968.
Unesco Survey, n.s. 3351/66.

6.095 GIUSTI, Gino Paul
'An economic study on research and development in the United
States since World War II'. Journal of Finance, Vol.15(1),
March 1960, pp.79-80.

6.096 GLOSKEY, Carl R.
'Research on a research department: an analysis of economic
decisions on projects'. IRE Transactions on Engineering
Management, 1960, Vol.EM-7(4), 166-172.

6.097 GOLD, Bela
Research, technological change and economic analysis.
Lexington, Mass.: Lexington Books, 1977. 240p

6.098 GOLD, Bela
Productivity, technology and capital: economic analysis,
managerial strategies, and government policies. Lexington,
Mass.: Lexington Books, 1979. 318p.

6.099 GOLD, Bela and KRAUS, Ralph M.

'Integrating physical with financial measures for managerial controls'. Journal of the Academy of Management, 1964, Vol. 7(2), 109-127.

6.100 GORDON, Gerald and others
'Organization for scientific productivity'. American Behavioral Scientist. Vol.5(4), December 1969, pp.35-37.

6.101 GRANDSTRAND, Ove
Technology management and markets: an investigation of R & D and innovation in industrial organizations. Gotenborg: Svenska Kulturkompanist, 1979. 417p.

6.102 GREAT BRITAIN - Select Committee on Science and Technology, Session 1977-78
Innovation research and development in Japanese science based industry. Vol.1: report. London: HMSO, 1978. (HX 682-1).

6.103 GREGORY, Carl E.
Management of intelligence: scientific problem solving and creativity. New York, McGraw, 1967.

6.104 GRIFFITHS, Dorothy and PEARSON, Alan.
'The marketing of research and development'. Industrial Marketing Management, Vol.2, June 1973, p. 261-272.

6.105 GRILICHES, Zvi and SCHMOOKLER, Jacob
'Inventing and maximising'. American Economic Review, 1963, Vol.53(4), 725-729.

6.106 GROSSFIELD, K.
'Inventions as business'. Economic Journal, 1962, Vol.72 (85), 12-26.

6.107 GROSSFIELD, K.
'The exploitation of invention in Britain by NRDC'. In: Polytechnic School of Management Studies, Report on the proceedings of a seminar on the economics of research and development. pp. 20-29. 1967.

6.108 GRUBER, W.H. and NILES, J.S.
'Research and experience in management: the need for a synthesis'. Business Horizons, August 1973. p. 15-24.

6.109 GUSTAFSON, W.Eric
'Research and development, new products and productivity'. American Economic Review, 1962, Vol.52(2), 177-185.

6.110 HAGGERTY, P.E.

'Research Management - a survival issue'. IEEE Transactions on Engineering Management, March 1964, Vol. EM-11(1), 3-7.

6.111 HAGGERTY, Partick E.
'Management of innovation'. Science Journal, April 1970, Vol. 6(4), 75-78.

6.112 HAGSTROM, Warren O.
'Traditional and modern forms of scientific teamwork'. Administrative Science Quarterly, 1964, Vol. 9(3), 241-263.

6.113 HAINE, R.W. and LOB, W.
'The application of closed-loop techniques to engineering project planning'. IRE Transactions on Engineering Management, 1960, Vol. EM-7(3), 96-103.

6.114 HAINER, Raymond M., KINGSBURY, Sherman and GLEICHER,
        David (eds)
Uncertainty in research management and new product development. New York, London: Reinhol. 1967. 234p.

6.115 HALL, Douglas T. and LAWLER, Edward E.
'Unused potential in research and development organisations'. Research Management, September 1969, Vol.12(5), 339-354.

6.116 HALL, Marshall
'The determinants of investment variations in research and development'. IEEE Transactions on Engineering Management, March 1964, Vol. EM-11(1), 8-15.

6.117 HAMBERG, Daniel
'Invention in the industrial research laboratory'. Journal of Political Economy, 1963, Vol.71(2), 95-115.

6.118 HAMBERG, Daniel
R & D: essays on the economics of research and development. New York, Random House, 1966.

6.119 HAMILTON, Carl W.
Optimal control of research and development expenditures. Cambridge, Mass., Massachusetts Institute of Technology, 1969. AD 702 918.

6.120 HARVARD BUSINESS REVIEW
R & D management series. Harvard Business Review. 2 Vols. Boston, Mass. No date.

6.121 HEBELER, Henry Koester
Cost and schedule uncertainty and prediction in development

projects. Submitted for the degree of Master of Science in Management at the Massachusetts Institute of Technology, June 1970. 176p.

6.122 HEGEDUS, Andras
'Management and scientific research', Social Science Information, 1968, Vol. 7(3), 7-14.

6.123 HENDRY, C.M.
'Getting it all together: integrating R & D with corporate objectives into a fluid, kinetic organization is essential to boost the R & D payoff'. Research Development, February 1973, p. 50-52.

6.124 HERSEY, Paul and BLANCHARD, Kenneth H.
'Managing research and development personnel: an application of leadership theory'. Research Management, September 1969, Vol. 12(5), 331-338.

6.125 HERSHEY, R.L.
'Finance and productivity in industrial research and development'. Research Management, July 1966.

6.126 HERTZ, David Bendal
The theory and practice of industrial research. New York, McGraw-Hill, 1950.

6.127 HERZOG, Arnold
Colleague networks, institutional roles and the international transfer of scientific knowledge: the case of Ireland. Submitted in partial fulfillment of the requirements for the degree of Doctor of Philosophy at the Massachusetts Institute of Technology, August, 1975. 238 leaves. Photocopy.

6.128 HESS, S.W.
'A dynamic programming approach to R & D budgeting and project selection'. IRE Transactions on Engineering Management, December 1962, Vol. EM-9, 170-179.

6.129 HIBBARD, W.R.
'Materials R & D: planning, programming, budgeting and measurement'. Transactions of the American Society for Metals, 1969, Vol. 62, 1031-1038.

6.130 HIGGINS, T.
Research planning and innovation: a study of success and failure in innovation and the implications for R & D management and choice. Dublin: Stationery Office, 1977. 283p.

6.131 HILL, Kenneth M. and others
'How much basic research is enough? A problem in resource allocation'. Long Range Planning, March 1969, Vol.1(3), 38-43.

6.132 HILL, Stephen C.
'A natural experiment on the influence of leadership behaviour patterns on scientific productivity'. IEEE Transactions on Engineering Management, Vol.EM-17, No. 1, February 1970, p. 10-20.

6.133 HISCOCKS, Edward S.
Laboratory Administration. London, MacMillan, 1958.

6.134 HOROWITZ, Ira
'Regression models for company expenditure on and returns from research and development'. IRE Transactions on Engineering Management 1960, Vol. EM-7(1), 8-13.

6.135 HOROWITZ, Ira
'Some comments on the future of industrial research'. Journal of Business, April 1962, Vol. 35(2), 210-211.

6.136 HOROWITZ, Ira
'Research inclinations of a Cournot oligopolist'. Review of Economic Studies, 1963, Vol.30(2), 128-130.

6.137 HOWE, J.D. and MCFETRIDGE, D.G.
'The determinants of R & D expenditures'. Canadian Journal of Economics, Vol. 9, No. 1, February 1976, p. 37-71.

6.138 HOWTON, F. William
'Work assignment and interpersonal relations in a research organization: some participant observations'. Administrative Science Quarterly, March 1963, Vol. 7, 502-520.

6.139 HUGGINS, Eric
Objective project management in research and development.
Paper presented at the International Management Symposium on Management and Economics in the Electronic Industry, Edinburgh 17-20 March, 1970. Slough: Urwick Technology Management Ltd., 1970. 13 leaves.

6.140 IMPERIAL CHEMICAL INDUSTRIES LTD.
Research opportunities. I.C.I., 1971. 45p.

6.141 INDUSTRIAL & COMMERCIAL TECHNIQUES LTD.
Economic forecasting in planning and research. By Gordon L. Spangler. London, 1968.

6.142   INDUSTRIAL RESEARCH AND DEVELOPMENT IN 1975
Trade and Industry, 6 April 1979. p.p. 32-36.

6.143   INSTITUTE OF PRODUCTION ENGINEERS
Current and future trends of manufacturing management and
technology in the U.K. Report by the Technical Policy Board.
London: I.P.E., 1980. 200p.

6.144   JACKSON, Thomas W. and SPURLOCK, Jack M.
Research and development management.   Homewood,   Illinois:
Dow Jones-Irwin, 1966.

6.145   JANTSCH, E.
Some aspects of potential economic implications of current
research and development trends in the energy field. Paris:
O.E.C.D., 1966. 126p. Mimeo.

6.146   JEWKES, John
Government and high technology.   Third   Wincott   Memorial
lecture   delivered   at   the   London   School of Economics and
Political Science, 31 October, 1972. London: Institute   of
Economic Affairs, 1972. 24p.

6.147   JOHNSTON, Ron
Contract research – a solution to what?   A   paper   presented
to   Australian   Industrial   Research Group, Sydney, November
1979. Wollongong, N.S.W.:   Department of HPS, University   of
Wollongong, 1979. 36 leaves. Mimeo. (Science, Technology and
Public Policy Report No.4).

6.148   JOHNSTON, Ron
Mechanisms for the co-ordination of research. Report to   the
Department   of   Science   and   the   Environment.   Wollongong,
N.S.W.:   University of Wollongong, 1979. 156p. Mimeo.

6.149   JOHNSTON, Ron
A role for the contract mechanism in Australian science and
technology.   Prepared as an Occasional Paper for the Science
Policy   Research   Centre,   School   of   Science,   Griffith
University,   Queensland,   March   1979.   Woolongong,   N.S.W.:
Department   of   HPS,   University   of   Wollongong,   1979.   36
leaves.   Mimeo.   (Science,   Technology   and   Public   Policy,
Report No.5).

6.150   JOLLY, J.A. and CREIGHTON, J.W. (eds)
Technology transfer in research and development. Proceedings
of the Briefing on Technology Transfer Projects organised by
the   Naval   Postgraduate   School   Monterey,   California.
Monterey: Naval Postgraduate School, 1975. 90p.

6.151 JOSEPH, R.A.
<u>A comparative study of Federal government incentive</u>
<u>programmes for industrial research and development in</u>
<u>Australia and Canada.</u> Australia: Griffith University, 1978.
89p. Mimeo.

6.152 KANAYAMA, Nobuo
'Towards a new dimension of R & D: an international
perspective'. <u>The Oriental Economist,</u> February 1971, pp.
24-36.

6.153 KANE, Henry
'Research as an aid to industrial development'. <u>State</u>
<u>Government (Oregon),</u> Spring 1962, 105-111.

6.154 KAPLAN, Norman
'Some organizational factors affecting creativity'. <u>IRE</u>
<u>Transactions,</u> March 1960, Vol. EM-7(1), 24-30.

6.155 KAPLAN, Norman
'Research administration and the administrator: U.S.S.R.
and U.S.'. <u>Administrative Science Quarterly,</u> June 1961, Vol.
6(1), 51-72.

6.156 KARGER, Delmar W.
<u>The new product: how to find, test, develop, cost, price,</u>
<u>protect, advertise and sell new products.</u> New York, Indus-
trial Press, 1960.

6.157 KATZ, R. and ALLEN, T.J.
'Investigating the Not Invented Here (NIH) syndrome: a look
at the performance tenure, and communication patterns of 50
R & D project groups'. <u>R & D Management,</u> Vol. 12, No. 1,
January 1982, p. 7-19.

6.158 KAY, Neil M.
<u>Corporate decision making for allocations to research and</u>
<u>development.</u> Strathclyde: Strathclyde University. No date.
30 leaves.

6.159 KAY, Neil M.
<u>Towards an open systems framework for the analysis of</u>
<u>research and development and innovative activity.</u> Stirling:
University of Stirling, 1974. 53 leaves. (Discussion paper
no. 20)

6.160 KAY, Neil M.
<u>The allocation of resources to research and development in</u>
<u>the firm.</u> Thesis submitted to the University of Stirling for
the Degree of Doctor of Philosophy, 1976. Not paginated. 2

parts.

6.161 KAY, Neil M. and FREEMAN, Christopher
The innovating firm: a behavioural theory of corporate R&D.
Foreward by Christopher Freeman. London: MacMillan, 1979.
266p.

6.162 KAYSEN, Carl
'Improving the efficiency of military research and
development'. Public Policy, 1963, p. 219-273.

6.163 KEGAN, Daniel L.
A technology flow diagram. Evanston Ill. : Northwestern
University, 1966. 3p. (Program of Research on the Management
of Research and Development.)

6.164 KEGAN, Daniel L.
'Organizational development: description, issues, and some
research results'. Academy of Management Journal, December
1971, p. 454-464. (Program of Research on the Management of
Research and Development - Project 12).

6.165 KEGAN, Daniel L.
Trust, openness, and organisational development: short-term
relationships in research and development laboratories and a
design for investigating long-term effects. Evanston, Ill.:
Northwestern University, 1971. 287p. (Program of Research on
the Management of Research and Development - Project 12).

6.166 KEGAN, Daniel L. and RUBENSTEIN, Albert H.
'Measures of trust and openness'. Comparative Group Studies,
May 1972, p. 179-201. (Program of Research on the Management
of Research & Development - Project 13).

6.167 KIEFER, D.M.
'Winds of change in industrial chemical research'. Chemical
Engineering News, March 1964, Vol. 42, 88-109.

6.168 KISTIAKOWSKY, George D.
'Basic research: an industrial responsibility'. Research
Management, 1960, Vol. 3(2), 69-76.

6.169 KOIDE, Konosuke
Roles of research and development and of imitation in
technological progress. Paper presented at the international
seminar on Research and Development Planning, Tokyo, 2-6
April (no year). Tokyo: Council for Industry Planning. 23p.

6.170 KOOPMAN, Georg
R & D options concerning future problems of the European

car industry. Study prepared for the EC Commission – FAST DG XII. Brussels: CEC, 1981. 69p. and summary. Mimeo. (FAST Contract No. 6) (XII/432/81/EN).

6.171 KORNAI, Janos
Anti-equilibrium: on economic systems theory and the tasks of research. Amsterdam, Oxford: North-Holland, 1971. 402p.

6.172 KROEGER, William J.
'Can prophets yield profits in R & D management'. Army Research and Development News Magazine. October 1966, 36–37.

6.173 LAGERMALM, Gosta
The Swedish pattern for applied research and development – and the role of the Swedish Board for Technical Development. Stockholm: S.T.U., 1975. 28p.

6.174 LANTOS, Peter R.
'What R & D can expect from marketing research'. Chemtech, October 1973, p. 588–91.

6.175 LEACH, Rodney
'Determining the 'shape' of your R & D organization'. European Business, October 1969, p. 9–17.

6.176 LEVER, Brian George.
Planning technological change: a case study of the agricultural chemicals industry. Thesis presented in fulfilment of the requirements for the degree of Doctor of Philosophy in the Science Policy Research Unit of the University of Sussex, May 1978. Brighton: SPRU/Haslemere: ICI, 1978. 317 leaves.

6.177 LOCK, Dennis
Project management. London: Gower Press, 1968.

6.178 LOCKE, Brian
Investment and planning in R & D. Paper for the British Institute of Management, 10 April 1972. 26 leaves. Mimeo.

6.179 LOCKE, H.Brian.
The use of research in industry. Mimeo. Paper to be presented at the Symposium on 'The R & D Business', 22nd April 1970.

6.180 LOCKETT, A.G. and GEAR, A.E.
'Programme selection in research and development'. Management Science, June 1972, Vol. 18, No. 10, p. B-575 – B-590.

6.181   LOCKETT, A.G. and GEAR, A.E.
'Representation and analysis of multi-stage problems in R & D'. Management Science, Vol. 19, No. 8, April 1973, p. 947-960.

6.182   LORENZ, Christopher
Investing in success: how to profit from design and innovation. London: The Anglo-German Foundation for the Study of Industrial Society, 1979. 26p.

6.183   LOTHROP, Warren C.
Management uses of research and development. New York, Harper, 1964.

6.184   MAISONROUGE, J.G.
Technology, competitiveness and people: the challenge from within. Remarks made at the British Institute of Personnel Management, Harrogate, Yorkshire, Wednesday, October 22, 1980. Harrogate: British Institute of Personnel Management, 1980. 26 leaves. Mimeo.

6.185   MANCHESTER BUSINESS SCHOOL
Industrial R & D strategy and management - a challenge for the 1980's. 10th Anniversary Conference held on 30 June - 2 July 1980 at Manchester Business School. Editor: A W. Pearson. Manchester: Manchester Business School, 1980.

6.186   MANSFIELD, Edwin
'Contribution of R & D to economic growth in the United States'. Science, Vol.173, No.4021, pp.477-86, 4 February 1972.

6.187   MANSFIELD, Edwin
'Basic research and productivity increase in manufacturing'. American Economic Review, Vol. 70, No. 5, December 1980. p. 863-873.

6.188   MANSFIELD, Edwin
'How economists see R & D.' Harvard Business Review, Vol. 59, No. 6, November/December 1981, p. 98-106.

6.189   MANSFIELD, Edwin and BRANDENBURG, Richard
'The allocation, characteristics, and outcome of the firm's research and development portfolio: a case study'. Journal of Business, October 1966, Vol. 39(4), 447-464.

6.190   MANSFIELD, Edwin, TEECE, David and ROMEO, Anthony
Overseas research and development by U.S.-based firms. No imprint, no date. 19p. Mimeo. Added to stock June 1980.

6.191 MARSTRAND, Pauline
'Reorganization of Government R & D in Britain: an
opportunity missed'. Science Policy, Vol.1, No.6, pp.1-1 -
1-4, November/December, 1972.

6.192 MARTINO, R. L.
Project management. New York: Gordon & Breach, 1968.

6.193 MAXWELL, Philip
Learning and technical change in the steel plant of Acindar
S.A. in Rosario, Argentina. Buenos Aires: Banco Inter-
americano de Dessarrollo, 1976. 200p.

6.194 MAXWELL, Philip
Implicit R & D strategy and investment-linked R & D: a
study of the R & D Programme of the Argentine Steel firm
Acindar S.A. Buenos Aires: IDB/ECLA Research Programme in
Science and Technology, 1979. 35p.

6.195 MAXWELL, Philip
Technical and organizational change in steelplants: an
Argentine and a Brasilian Case. Simposio de Analisis
Organizational, Buenos Aires, 16-18 de Octubre de 1980.
Buenos Aires: Institute Torcuato De Tella, 1980. 20 leaves.
Mimeo.

6.196 MEADOWS, Dennis L. and MARQUIS, Donald G.
Characteristics and implications of forecasting errors in
the selection of R & D projects. Cambridge, Mass: Sloan
School of Management, 1968. (Research Program on the
Management of Science and Technology, paper 313-68.)

6.197 MEERSCH, R. Vander
Technological innovation, its nature and management.
Cranfield: Institute of Technology, 1970. 113 leaves.
Mimeograph.

6.198 MENTON, David C.
'Changing patterns in research', Battelle Technical Review,
July 1964.

6.199 MILLER, Roger Emile
Innovation, organization and environment: a study of
sixteen American and West European steel firms. Sherbrooke:
Institut en Recherche et de perfectionnement en
Administration, 1971. 211p.

6.200 MONTEITH, G. S.
R & D administration. London, Iliffe, 1969.

6.201 MORRISON, J. K.
'Controlling research', <u>Management Decision</u>, Summer 1968, Vol. 2(2).

6.202 MORTON, Jack A.
'From research to technology', <u>International Science and Technology</u>, May 1964, No. 29, 82-92.

6.203 MOTTLEY, Charles M. and NEWTON, R. D.
'The selection of projects for industrial research', <u>Operations Research</u>, 1959, Vol. 7, 740-751.

6.204 MOWERY, David C.
<u>The nature of the firm and the organisation of research: an investigation of the relationship between contract and in-house research.</u> Boston, Mass.:   Harvard   University, Graduate   School   of   Business   Administration,   1982.   34p. Mimeo. Working Paper HBS 82-40)

6.205 MUSE, William V. and KEGERREIS, Robert J.
'Technological   innovation   and   marketing   management: implications   for   corporate   policy', <u>Journal of Marketing</u>, Vol. 33, p. 3-9, October 1969.

6.206 McCULLOCH, D.G.
<u>Technology and innovation in New Zealand manufacturing industry - the role of government.</u> Resource   Paper 2. Lower Hutt, New Zealand: Physics   and   Engineering   Laboratory, DSIR, 1980. 66p. Mimeo. (PEL Report No.686).

6.207 McEACHERN, William A. and ROMEO, Anthony A.
'Stockholder control, uncertainty and the allocation of resources   to   research   and   development'. <u>Journal of Industrial Economics</u>,   Vol. XXVI,   No.   4,   June   1978,   p. 349-361.

6.208 McLEOD, T.S.
<u>Management of research, development and design in industry.</u> London, Gower Press, 1969.

6.209 McNEIL, P.C.
'R & D and industrial innovation'. <u>Proceedings of the Institution of Electrical Engineers</u>,   Vol.   126,   No.   1, January 1979, p. 117-122.

6.210 NASLUND, Bertil and SELLSTEDT, Bo
<u>Budgets for research and development: an empirical study of 69 Swedish firms.</u> Brussels: European Institute for Advanced Studies in Management, 1973. 24p. Working paper 73-7.

6.211 NASLUND, Bertil and SELLSTEDT, Bo
An evaluation of some methods for determining the R & D budget. Brussels: European Institute for Advanced Studies in Management, 1973. 21p. Working paper, 73-3.

6.212 NELSON, Richard R.
'Uncertainty, learning, and the economics of parallel research and development efforts'. Review of Economics and Statistics, 1961, 43(4), 351-364.

6.213 NIEMANN, Ralph A.
'Pitfalls in R & D Management'. Personnel, Jan-Feb, 1970, 46-51.

6.214 NOLTINGK, B.E.
The human element in research management. London, Elsevier, 1959.

6.215 NORMAN, Colin
Knowledge and power: the global research and development budget. Washington, D.C. Worldwatch Institute, 1979. (Worldwatch Paper 31). 56p.

6.216 NORWAY - Ministry of Education
The organization of research in Norway. Oslo: Ministry of Education, 1980. 9 leaves. Mimeo.

6.217 NYSTROM, Harry and EDVARDSSON, Bo
Research and development strategies for four Swedish farm machine companies. Uppsala, Sweden: Innovation Research Group, Institute for Economics and Statistics, 1980. 31 leaves. Mimeo.

6.218 NYSTROM, Harry and EDVARDSSON, Bo
Technological and marketing strategies for product development: a study of 20 Swedish food processing companies. Uppsala, Sweden: Swedish University of Agricultural Sciences, Department of Economics and Statistics, 1980. 167p and appendices. (Report No. 164).

6.219 OLIN, J.
R & D Management practices: chemistry industry in Europe. Final draft. November, 1972. Zurich: Stanford Research Institute, 1972. 118p.

6.220 ORTH, Charles D. and others
Administering research and development: the behaviour of scientists and engineers in organisations. London, Tavi-stock, 1965.

6.221 OSHIMA, Keichi
Research and development and economic growth in Japan.
Tokyo: Department of Nuclear Engineering, University of
Tokyo. No date. 19 leaves. Mimeo.

6.222 PARKER, R.C.
'The art and science of selecting and solving research and
development   problems: Automobile   Division Chairman's
address'. 12th October, 1971. Proc. Institution of
Mechanical Engineers, 1970-71, Vol.185, 64/71, p.879-893.

6.223 PARKER, R.C.
The encouragement of creative thinking in an engineering R&D
laboratory. Chapel-en-le Frith: Ferodo Ltd., 1974. 25
leaves. Mimeo.

6.224 PARKER, R.C.
R & D evaluation and selection. Paper presented at a
Conference on General Management Skills organised by the
Institution of Mechanical Engineers, Manufacture and
Management Group, held at Blackpool, 17-19 May, 1974.
London: IME, 1974. 19 leaves. Mimeo. (Conference
Publication 12).

6.225 PARKER, R.C.
'Human aspects of R & D organization'. Proc. of the
Institution of Mechanical Engineers, Vol. 189, No.42, 1975,
p.287-294.

6.226 PATERSON, T.T.
'Administration of research'. Nature, 1963, 198(4880),
520-525.

6.227 PAVITT, Keith
'Industrial R & D and the British economic problem'. R & D
Management, Vol.10, Special Issue, 1980. pp.149-158.

6.228 PEARLMAN, Jerome
'Engineering program planning and control through the use of
PERT'. IRE Transactions on Engineering Management. 1960.
EM-7(4), 125-134.

6.229 PEARSON, A.W.
The organisation of research centres. Paper to be presented
to the conference on strategies for management, research and
development in Europe, September 18-20, 1972. 16 leaves.
Mimeo.

6.230 PEARSON, A.W.
'R & D as a mini-business'. European Business, No. 38,

Summer 1973, pp.36-42.

6.231 PEIRSON, D.E.H.
'The organisation and planning of research and development'.
Atom, No.164, June 1970, pp.117-119.

6.232 PELZ, Donald C.
Time and influence factors in laboratory management, as
related to performance. Ann Arbor, University of Michigan
Survey Research Center, 1962.

6.233 PELZ, Donald C.
'Freedom in research'. International Science and Technology,
1964. No.26, 54-66.

6.234 PELZ, Donald C. and ANDREWS, Frank M.
'Organizational atmosphere, motivation, and research
contribution'. American Behavioral Scientist, 1962, 6(4),
pp.43-47.

6.235 PELZ, Donald C. and ANDREWS, Frank M.
'Diversity in research'. International Science and
Technology, 1964. No.31, pp.28-36.

6.236 PELZ, Donald C. and ANDREWS, Frank M.
Scientists in organizations: productive climates for
research and development. New York, Wiley, 1966.

6.237 PETERS, Donald H.
Commercial innovations from university faculty: a study of
the invention and exploitation of ideas.   Cambridge,
Mass.:M.I.T., 1969. 28 leaves. (Research program on the
management of science and technology).

6.238 PETERS, Donald H. and ROBERTS, Edward B.
Unutilized ideas in university laboratories: a preliminary
study. Cambridge, Mass., Sloan School of Management, 1967.
Research Program on the Management of Science and
Technology, paper 287/67.

6.239 POUST, Roy and RUBIN, Irwin
Analysis of problems encountered in R & D project
management. Cambridge, Mass., Sloan School of Management,
1966. Research Program on the Management of Science and
Technology, paper 232-66.

6.240 PRUTHI, S. and NAGPAUL, P.S.
Pattern and role of informal communication in R & D.   New
Delhi: Council of Scientific and Industrial Research, 1978.
49p. (Monograph/Study Report Series - 1/78).

6.241 QUINN, James Brian
'Long-range planning of industrial research'. <u>Harvard</u>
<u>Business Review</u>, July - August 1961, 39, 88-102.

6.242 QUINN, James Brian and MUELLER, James A.
'Transferring research results to operations'. <u>Harvard</u>
<u>Business Review</u>, 1963, 41(1), 49-66.

6.243 RADNOR, Michael and NEAL, Rodney
<u>The progress of management science activities in large U.S.</u>
<u>industrial corporations.</u> Evanston, Illinois: Northwestern
University, 1971. 33 leaves. (Program of research on the
management of Research and Development - Project 7). Mimeo.

6.244 RAFFAELE, Joseph Antonio
<u>The management of technology: change in a society of</u>
<u>organized advocacies.</u> Revised edition. Washington, D.C.:
University Press of America, 1979. 346p.

6.245 RAMAER, J.C.
'Know-how and its transfer - do we know how? <u>Industrial</u>
<u>Research and Development News</u> 1967, 2(1), 13-15.

6.246 RANDLE, C. Wilson
'Problems of R & D Management'. <u>Harvard Business Review</u>,
January-February 1959, 37(1), 128-136.

6.247 REICH, Irving
'Creativity in research organisations'. <u>Research Management</u>,
1960, 3(4), 217-226.

6.248 RICHARDSON, Jacques (ed)
<u>Integrated technology transfer.</u> Mount Airey, Maryland:
Lomond Books, 1979. 162p.

6.249 RILEY, Derek
'Financing transportation innovation in perspective'.
<u>Aeronautical Journal</u>, 77, No.748, April 1973, pp.181-189.

6.250 RIVKIN, Steven R.
<u>Technology unbound: transferring scientific and engineering</u>
<u>resources from defense to civilian purposes.</u>   Oxford,
Pergamon, 1968.

6.251 ROBERTS, Edward B.
<u>The dynamics of research and development.</u> New York, Harper,
1964.

6.252 ROBERTS, Edward B.
'The myths of research management'. <u>Science and Technology</u>,

August 1968, No.80, 40-46.

6.253 ROBERTSON, A. and ROTHWELL, R.
'The contribution of poor communications to innovative failure'. Aslib Proc. 27, (10), October 1975. pp.393-400.

6.254 ROMAN, Daniel David
Research and development management: the economics and administration of technology. New York, Appleton-Century, 1968.

6.255 RONAYNE, Jarlath
The allocation of resources to research and development - a review of policies and procedures. Report to the Australian Science and Technology Council. Sydney: University of New South Wales. No date. 132 leaves. Mimeo. Added to stock, July 1980.

6.256 RONSTADT, Robert
Research and development abroad by U.S. multinationals. New York: Praeger, 1977. 127p.

6.257 ROSEN, E.M. and SOUDER, W.E.
'A method for allocating R & D expenditures'. IEEE Transactions on Engineering Management, Vol.EM-12, September 1965, pp.87-93.

6.258 ROTHWELL, Roy
Marketing and successful industrial innovation. Marketing Theory Seminar, Innovation in Marketing. May 1973. 32 leaves. Mimeo.

6.259 ROTHWELL, Roy
'Nucleonic thickness gauges - a SAPPHO pair'. Research Policy 2, 1973, 144-156p.

6.260 ROTHWELL, Roy
'Factors for success in industrial innovation'. Journal of General Management, Vol.2, No.2, Winter 1975, pp.57-65.

6.261 ROTHWELL, Roy
'From invention to new business via the New Venture Approach'. Management Decision, 13(1), 1975, pp.10-21.

6.262 ROTHWELL, Roy
'Stibbe-Monk: case study holds salutary lessons for technical management'. Knitting World and Textile Manufacturer, ITMA Review Issue, Vol.2, No.3, 1975, pp.10-12, 27.

6.263 ROTHWELL, Roy
'Stibbe—Monk story - 2. Patternmaster: project that flopped for lack of a "product champion"'. Knitting World & Textile Manufacturer, No.1, 1976, pp.21-23.

6.264 ROTHWELL, Roy
Innovation in textiles: some significant factors in success and failure. Brighton: SPRU, University of Sussex, 1976. 44p. (SPRU Occasional Paper Series, No.2).

6.265 ROTHWELL, Roy
'Innovation in the U.K. textile machinery industry: the results of a postal questionnaire survey'. R & D Management, Vol.6, No.3, June 1976, pp.131-8.

6.266 ROTHWELL, Roy
'Intracorporate entrepreneurs'. Management Decision, 13(3), 1976, pp.142-154.

6.267 ROTHWELL, Roy
'The UK textile machinery industry: a case study in technical change'. Published in: BOWE G. (ed) Industrial efficiency and the role of government. HMSO, 1977. pp.137-169.

6.268 ROTHWELL, Roy
'Users' and producers' perceptions of the relative importance of various textile machinery characteristics'. Textile Institute and Industry, July 1977, pp.239-242.

6.269 ROTHWELL, Roy
'The characteristics of successful innovators and technically progressive firms (with some comments on innovation research)'. R & D Management, Vol.7, No.3, June 1977, pp.191-206.

6.270 ROTHWELL, Roy
Innovation in textile machinery and its relevance to the textile industry. Prepared for MIRA Conference 'Managing for survival and profitability', Leeds, 28 October, 1977. 43 leaves. Mimeo.

6.271 ROTHWELL, Roy
'The management of textile machinery innovation: some lessons of failure'. Textile Institute and Industry, April 1977. 130-134p.

6.272 ROTHWELL, Roy
'Marketing - a success factor in industrial innovation'. Management Decision, 14(1), pp.43-53.

6.273 ROTHWELL, Roy
'Some problems of technology transfer into industry:
examples from the textile machinery sector'. IEEE
Transactions on Engineering Management Vol.EM-25, No.1,
February 1978, pp.15-20.

6.274 ROTHWELL, Roy
'Where Britain lags behind - competition'. Management Today,
November 1978, pp.82-85.

6.275 ROTHWELL, Roy
'Why British farm machinery is losing markets to the
Continentals'. Agricultural Machinery Journal, October 1979,
pp.26-29.

6.276 ROTHWELL, Roy
'It's not (just) what you make, it's the way you make it'.
Design, March 1980, No.375, pp.44-45.

6.277 ROTHWELL, Roy
Some developments in innovation research methodology:
results for the agricultural engineering industry. Paper
prepared for the International Workshop on 'Innovation for
efficiency', IIASA, Schloss Laxenburg, Austria, June 22-25,
1981. Brighton: SPRU, April 1981. 22 leaves. Mimeo.

6.278 ROTHWELL, Roy
'The market factor in innovation: some lessons of failure'.
The Specialist (International), Vol.2, Paper No.5,
July/August 1982, pp.9-14.

6.279 ROTHWELL, Roy
'The commercialization of university research'. Brighton:
SPRU, May 1982. 22 leaves. Mimeo. Submitted to Physics
in Technology, November 1982.

6.280 ROTHWELL, Roy
Innovation as a local phenomenon: creating an innovation
infrastructure. Prepared for: SEFI Conference 'The
education of the Engineer for innovative and entrepreneurial
activity', June 23-25, 1982. Published by Delft University
Press. Brighton: SPRU, March 1982. 9 leaves. Mimeo.

6.281 ROTHWELL, Roy and ROBERTSON, A.
'The role of communications in technological innovation'.
Research Policy Vol.2, No.3, October 1973, pp.204-225.

6.282 ROTHWELL, Roy and TEUBAL, Morris
'SAPPHO revisited: a re-appraisal of the SAPPHO data'. In:
STROETMANN, Karl A. (ed) Innovation, economic change and

technology policies. Proceedings of a seminar on Technological Innovation, Bonn, 5-9 April, 1976. Basel: Birkhauser Verlag, 1977. pp.39-59.

6.283 ROTHWELL, Roy et al
'SAPPHO updated - project SAPPHO phase II'. Research Policy, Vol.3, No.3, pp.258-291, November 1974.

6.284 RUBENSTEIN, Albert H.
'Setting criteria for R & D'. Harvard Business Review 35(1), 95-104, 1957.

6.285 RUBENSTEIN, Albert H.
'Organization and research and development decision making within the decentralized form'. In: NATIONAL BUREAU OF ECONOMIC RESEARCH The rate and direction of inventive activity: economic and social factors. 1962.

6.286 RUBENSTEIN, Albert H.
Methodological aspects of in-house studies of R & D management. Typescript, Program of Research on the Management of Research and Development, 1964. Presented at the 11th international meeting, Institute of Management Science, 1964.

6.287 RUBENSTEIN, Albert H.
'Organisational factors affecting research and development decision-making in large decentralized companies'. Management Science, Vol.10, No.4, July 1964. pp.618-633.

6.288 RUBENSTEIN, Albert H.
Some observations on research-on-research in Europe. Typescript. Evanston, Illinois, Northwestern University, Program of Research on the Management of Research and Development, 1969.

6.289 RUBENSTEIN, Albert H.
A study of international technology transfer - collaborative research between Northwestern University and the University of Cologne, West Germany. Mimeo. Program of Research on the Management of Research and Development, Northwestern University, 1969.

6.290 RUBENSTEIN, Albert H.
Coupling relations in research and development. (Annual progress report 1970-1971). Evanston, Illinois: Northwestern University, 1971. 19 leaves. (Program of research on the management of Research & Development - Project 8 (2)). Mimeo.

6.291 RUBENSTEIN, Albert H.
Final report: studies and analyses of the management of
scientific research and development, including
implementation and application at NASA centers. Evanston,
Illinois: Program of research on the Management of Research
and Development, 1975. 20p.

6.292 RUBENSTEIN, Albert H. and HANNENBERG, Richard C.
Idea flow and project selection in several industrial
research and development laboratories. Prepared for the
Conference on Economic and Social Factors in Technological
Research and Development, Ohio State University, October
15-17, 1962.

6.293 RUBENSTEIN, Albert H., BARTH, Richard T. and DOUDS,
       Charles F.
'Ways to improve communications between R & D groups'.
Research Management, November 1971, pp.49-59. (Program of
Research on the Management of Research & Development -
Project 8 (1)).

6.294 RUPP, Erik
'The RKW: a new approach towards technology transfer.
Methods for the promotion of innovation in small- and
medium-sized companies'. Research Policy,   Vol.5,   No.4,
October 1976, pp.398-412.

6.295 SAINT-ROSSY, Dan T., CHAKRABARTI, Alok K. and
       RUBENSTEIN, Albert H.
International transfer of technology: an evaluation of the
education & training component. Evanston, Illinois: Program
of Research on the Management of Research & Development,
1976. 28p.

6.296 SANDKULL, Bengt
On product changes and product planning. Lund,   Swedish
Institute for Administrative Research, 1968.

6.297 SAPPHO
Success and failure in industrial innovation. Report   on
project Sappho by the Science Policy Research Unit,
University of Sussex. London: Centre for the Study of
Industrial Innovation, 1972. 36p.

6.298 SAVEREIDE, Thomas J.
'A case history - stimulating innovation in industry'.
Research/Development, November 1973, pp.22-24.

6.299 SCHERER, F.M.
The structure of industrial technology flows.   Chicago:

Northwestern University, 1981, 22 leaves. Mimeo.

6.300 SCHNITTGER, Jan R.
'Research and development in the Swedish engineering industry today'. In: European conference on technological forecasting, Glasgow, June 24-26, 1968. Typescripts, 1968.

6.301 SCHOLES, D.H.C.
'Getting value for money from R & D'. Research 2(3), May 1969, pp.4-13.

6.302 SCHOTT, Kerry
Industrial research and development expenditure: an econometric analysis. Thesis submitted for the Degree of Doctor of Philosophy at the University of Oxford, July 1975. 267 leaves.

6.303 SCHOTT, Kerry
'The relations between industrial research and development and factor demands'. The Economic Journal 88, March 1978, pp.85-106.

6.304 SCHRAGE, Harry
'The R & D entrepreneur: profile of success'. In: Harvard Business Review. R & D Management series.... Vol.2. 1965.

6.305 SCHRIER, Elliot
'Toward technology transfer'. Technology and Culture, 5(3), Summer 1964, pp.344-358.

6.306 SEILER, Robert E.
Improving the effectiveness of research and development. New York, McGraw-Hill, 1965.

6.307 SELLSTEDT, Bo and NASLUND, Bertil
Product development plans. Brussels: European Institute for Advanced Studies in Management, 1972. 22 leaves.

6.308 SHARPLES, Allan
'The role of independent R & D organisations'. Chemistry and Industry, 18 August 1973, pp.774-777.

6.309 SHEDDEN, W.I.H.
'The global deployment of research and development resources'. Chemistry and Industry, 2nd June 1973, pp.509-510.

6.310 SHEPARD, Herbert A.
'Patterns of organization for applied research and development'. Journal of Business of the University of

Chicago,3 29(1), 1956, pp.52-58.

6.311 SIMMONDS, W.H.C.
'Planning and R & D in a turbulent environment'. Research Management, November 1975, pp.17-21.

6.312 SIPRI
Resources devoted to military research and development: an international comparison. Stockholm: Almqvist and Wiksell, and London: Paul Elek, 1972. 112p. Pbk.

6.313 SOLO, Robert A.
'Research and development in the synthetic rubber industry'. Quarterly Journal of Economics, 68(1), February 1954, pp.61-82.

6.314 SOMASEKHARA, N.
'Some management aspects of R & D in industry'. Indian Management, September 1972, pp.1-5.

6.315 SOUDER, William E.
'Analytical effectiveness of mathematical models for R & D project selection'. Management Science, Vol.19, No.8, April 1973, pp.907-923.

6.316 SOUDER, William E.
An exploratory study of the coordinating mechanisms between R & D and marketing as an influence on the innovation process. Final report, with contributions by A.K. Chakrabarti and T.V. Bonoma. Pittsburgh: Technology Management Studies Group, University of Pittsburgh, 1977. Not paginated.

6.317 SOUDER, William E., MAHER, P. Michael and RUBENSTEIN, A.H.
'Two successful experiments in project selection'. Research Management, Vol.15, No.5, September 1972, pp.44-54.

6.318 STANKIEWICZ, Rikard
'The effects of leadership on the relationship between the size of research groups and their scientific performance'. R & D Management, Vol.9, Special Issue, 1979, pp.1-13. (University of Lund, Research Policy Institute Offprint Series No.28).

6.319 STANKIEWICZ, Rikard
Leadership and the performance of research groups. Lund: University of Lund, Research Policy Institute, 1980. 143p.

6.320 STEELE, Lowell W.

Incentives for technological innovation. Schenectady, New York: Corporate Research and Development, G.E.C., 1977. 30 leaves. Mimeo.

6.321 SUGDEN, T. Morris and AFFLECK, William S.
'Role of companies R & D organisations'. Chemistry and Industry, 18 August 1973, pp.769-773.

6.322 SZAKASITS, Georges D.
'The adoption of the Sappho method in the Hungarian electronics industry'. Research Policy, Vol.3, No.2, April 1974, pp.18-28.

6.323 TERLECKYJ, Nestor E. and HALPER, H.J.
Research and Development: its growth and composition. New York, National Industrial Conference Board, 1963.

6.324 THE RESEARCH AND DEVELOPMENT SOCIETY
In Europe now: co-operation in research and technology.
London: The Research & Development Society, 1973. 162p.

6.325 THE RESEARCH AND DEVELOPMENT SOCIETY
Whither R & D? Proceedings of the 1976 symposium held on 27th April at the Royal Society, London. London: The Research and Development Society, 1976. 73p.

6.326 THOMAS, George F.
The management of research and development. London,   Batsford, 1970.

6.327 TWISS, Brian C.
Strategy for research and development. Bradford: University of Bradford Management Centre. No date. 13 leaves. Mimeo.

6.328 TYBOUT, Richard A. (ed)
Economics of research and development. Ohio   State   U.P., 1965.

6.329 UNITED STATES - National Science Foundation
Support of basic research by industry. By Howard K. Nason, Joseph A. Steger and George E. Manners. Washington, D.C.: N.S.F., 1978. 55p.

6.330 UNITED STATES - National Science Foundation
How basic research reaps unexpected rewards.   Washington D.C.: National Science Foundation, 1980. 37p.

6.331 VEDIN, Bengt-Arne (ed)
Current innovation: policy, management and research options.
Stockholm: Almquist and Wiksell International, 1980. 203p.

6.332 VON HIPPEL, Eric
A customer-active paradigm for industrial product idea
generation. Cambridge, Mass.: Alfred P. Sloan School of
Management, 1977. 37 leaves. (Working paper 935-77).

6.333 WILSON, R.M.S.
'Applying financial control to R & D'. Management
Accounting, May 1973, pp.199-202.

6.334 WOODWARD, F.
'How to manage research'. Management Today, September 1968,
pp.66-69.

6.335 WOODWARD, F.N.
'The strategy of R & D'. The Director, April 1968,
pp.106-108.

6.336 WORMALD, Avison
'Managing the scientist'. Management Today, February 1969,
pp.97-104.

6.b Project Evaluation In R & D

6.337 ALLEN, J.M. and NORRIS, K.P.
Project estimates and outcomes in research for a process
industry. Manchester: Manchester Business School. No date.
12 leaves. Mimeo. (Manchester Business School internal memo.
No.29).

6.338 BAER, Walter S., JOHNSON, Leland L. and MERROW,
       Edward W.
Analysis of Federally funded demonstration projects, Vol.2:
final report. Santa Monica, California: Rand, 1976. 188p.

6.339 BOSWORTH, Derek L. and WILSON, R A.
Returns to scale in research and development activity:
empirical evidence from the world chemical industry.
Coventry: Centre for Industrial Economic and Business
Research, University of Warwick, 1980. 22p. Mimeo. (General
Series Discussion Paper Number 89).

6.340 BROCKHOFF, Klaus
'On the quantification of the marginal productivity of
industrial research by estimating the production function
for a single firm'. The German Economic Review, 1970, 8(3),
202-229.

6.341 CAHALAN, M.J.
'R & D Project Evaluation'. Metals and Materials, February
1970, 55-57.

6.342 CENTRE FOR THE STUDY OF INDUSTRIAL INNOVATION
On the shelf: a survey of industrial R & D projects
abandoned for non-technical reasons. London: The Centre for
the Study of Industrial Innovation, 1971. 46p.

6.343 COMMISSION OF THE EUROPEAN COMMUNITIES
Evaluation of the Community's indirect action programme on
management and storage of radioactive waste. Brussels: CEC,
1981. 127p. Mimeo. (Research Evaluation-Report No. 4)

6.344 COMMISSION OF THE EUROPEAN COMMUNITIES
The evaluation of the Community Bureau of Reference
indirect action programme, 1975-1978. Brussels; CEC, 1982.
6p. Mimeo.

6.345 COMMISSION OF THE EUROPEAN COMMUNITIES
The evaluation of the Community's first energy R & D
programme (1975-1979). Brussels: CEC, 1982. 10p.

6.346 COMMISSION OF THE EUROPEAN COMMUNITIES
The evaluation of research and development. Conference held
on 25-26 January 1982, Brussels. Brussels: CEC, 1982.
Assorted papers. Mimeo.

6.347 COMMISSSION OF THE EUROPEAN COMMUNITIES
Evaluation of the concerted actions of the Community's
first medical research programme, 1978-1981. Brussels CEC,
1981. 36p. Mimeo. (Research Evaluation-Report No. 5).

6.348 COMPTON, E.
'Tools for R & D evaluation'. Financial Executive, February
1968, 31-40.

6.349 DEAN, Burton V. and NISHRY, Meir J.
'Scoring and profitability models for evaluating and
selecting engineering projects', Journal of the Operations
Research Society of America, Vol.13, No.4, July/August,
1965, p.550-570.

6.350 DEDIJER, Stevan and LONGRIGG, A.
'A model of foreign research policy'. Scientific World,
No.1, 1969, pp.17-21.

6.351 EUROPEAN ECONOMIC COMMISSION
Evaluation of Research and Development. Seminar held at
Copenhagen, 29 June - 1 July 1978. (No imprint), 1978.

(Assorted papers). Mimeo.

6.352 FABIAN, Y.
Measurement of output of R and D.   Paris,   OECD,   1963.
DAS/PD/63.48.

6.353 GERSTENFELD, Arthur
'A study of successful projects, unsuccessful projects and
projects in process in West Germany'. IEEE Transactions on
Engineering Management, Vol.EM-23,   No.3,   August   1976.
pp.116-123.

6.354 GREAT BRITAIN - Programmes Analysis Unit
A discussion of some methods for the comparative evaluation
of a set of projects. Chilton: P.A.U., 1974. 32p. (PAU
M26).

6.355 GREAT BRITAIN - Programmes Analysis Unit
Project appraisal and needs research - part 1 of collected
essays. Chilton: PAU, 1974. 99p.

6.356 HART, A.
'A chart for evaluating product research and development
projects'. Operational Research Quarterly, 1966, Vol.17(4),
347-358.

6.357 JONES, P.M.S.
Project evaluation - decisions on levels of investment in
research, feasibility studies and market surveys.   Chilton,
Didcot, Berks, Programmes Analysis Unit, 1969.

6.358 KARGER, Delmar W. and MURDICK, Robert G.
'The art and science of forecasting new product payoff:
here's how 37 companies practice'. Machine Design, 30
September 1971, Vol. 43, No. 24, p. 38-42.

6.359 LAWRENCE, J.R.
'Evaluating research'. Management Decision, Winter 1968,
Vol. 3(4), 241-248.

6.360 LENZ, Ralph C., MACHNIC, John A. and ELKINS, Anthony
       W.
The influence of aeronautical R & D expenditure upon the
productivity of air transportation.   Dayton,   Ohio:
University of Dayton Research Institute, 1981. 216p. &
Appendices. Mimeo. (UDR-TR-81-72)

6.361 LIBIK, George
The economic assessment of research and development: a
survey of recent practice in Western Europe, Eastern Europe

and the United States. Stockholm, Royal Institute of
Technology, Typescript, 1963.

6.362 LODGE, R.M.
'Choice and valuation in industrial research'. Advancement
of Science, 1958, Vol. 15(59), 211-216.

6.363 MAHER, Michael P.
Results from an experiment with a computer-based R & D
project selection technique. Edmonton, Alberta: University
of Alberta, 1972. 31 leaves. (Program of Research on the
Management of Research & Development - Project 10). Mimeo.

6.364 MARSTRAND, Pauline
'Biological effects as part of the evaluation of industrial
projects'. Biologist, 23(3), pp.140.

6.365 NICHOLSON, R.L.R.
Programme evaluation in the Programmes Analysis Unit.
Chilton, Berks., Programmes Analysis Unit, 1967.

6.366 NICHOLSON, R.L.R.
Evaluating civil research and development projects in the
government sector. Paper given at meeting of Institute of
Physics and Physical Society, Monday, 5th May, 1969.
Chilton, Didcot, Berks, Programmes Analysis Unit, 1969.

6.367 NORDEN, P.V.
'On the anatomy of development projects'. IRE Transactions
on Engineering Management, 1960, EM-7(1), 34-42.

6.368 NORTH ATLANTIC COUNCIL - Science Committee
Forum discussion on evaluation of research. Brussels: NATO,
1982. 11 leaves. Mimeo.

6.369 PARK, William R.
Cost engineering analysis: a guide to the economic
evaluation of engineering projects. New York, London: John
Wiley & Sons, 1973. 300p.

6.370 PEARSON, A.W. and ALLEN, J.M.
'Assessing research and development'. Science Journal,
August 1969, 5A(2), 79-83.

6.371 PEARSON, A.W. and TOPALIAN, A.S.
'Project evaluation in research and development'.
Management Decision, 3(3), Autumn 1969, pp.26-29.

6.372 POUND, William H.
'Research project selection: testing a model in the field'.

160

<u>IEEE Transactions on Engineering Management</u>,   March   1964,
EM-11(1), 16-22.

6.373   PRACTICAL CONCEPTS INCORPORATED
<u>The feasibility of monitoring expenditures for technolog-
ical innovation.</u> Vol. 1 - Highlights of the final report on
phase one. Vol. 2 - Final report on phase one. Washington,
D.C.: Practical Concepts Inc., 1974. 15p & 80p.

6.374   QUINN, James Brian
'How to evaluate research output'. <u>Harvard Business Review</u>,
March 1960, 69-80.

6.375   ROBERTSON, Andrew
'The innovators: profit or loss from industrial research?'
<u>Business Administration</u>, March 1969, 35-37.

6.376   RUBENSTEIN, Albert H.
'Economic evaluation of research and development: a brief
survey of   theory and practice'. <u>Journal of Industrial
Engineering</u>,Vol.17, No.11, November 1966. pp.615-620.

6.377   SCHWARTZ, Hugh and BERNEY, Richard (eds)
<u>Social and economic dimensions of project evaluation.</u>
Proceedings and papers of the Symposium on the Use of
Socioeconomic Investment Criteria in Project Evaluation held
at the Inter-American Development Bank on March 28-30, 1973.
Washington D.C.:   Inter-American Development Bank, 1977.
338p. Pbk.

6.378   SIMPSON, David
<u>The measurement and control of R & D projects.</u> Paper given
at the University of Sussex Research Seminar, March 13,
1970. 18 leaves.

6.379   SLOT, Wessel
<u>Project evaluation techniques in industrial research and
development: their use and limitations.</u>   University   of
Sussex, 1974. 11 leaves. Mimeo.

6.380   SOLOMON, Morris J.
<u>Analysis of projects for economic growth: an operational
system for their formulation, evaluation and implementation.</u>
New York, Praeger, 1970.

6.381   SPILLER, Pablo and TEUBAL, Morris
<u>Analysis of R & D failure.</u> Jerusalem: Maurice   Falk
Institute for Economic Research in Israel, 1975. 33p.
(Discussion paper No.7512).

6.c Technology Assessment

6.382 ALEXANDER, Arthur J. and NELSON, J.R.
Measuring technological change: aircraft turbine engines.
Santa Monica, Ca.:Rand, 1972. 37 leaves. (R-1017-ARPA/PR).

6.383 BROADWAY, Frank
'Why technology doesn't pay'. Management Today, January
1969, 52-57.

6.384 BROOKS, Harvey and BOWERS, Raymond
'The assessment of technology'. Scientific American, 222(2),
February 1970, pp.13-21.

6.385 CLAUSER, H.R.
Progress in assessing technological innovation - 1974.
Westport: Technomic, 1975. 103p.

6.386 COUNCIL FOR SCIENCE AND SOCIETY
Superstar technologies: the problem of monitoring
technologies in those instances where technical competence
is monopolised by a small number of institutions committed
to the same interest. London: Barry Rose (Publishers) Ltd.
in association with the Council for Science and Society,
1976. 66p.

6.387 DADDARIO, Emilio Q.
'Technology assessment'. Technology Review, December 1967,
Vol.70, 15-19.

6.388 FROST, Michael J.
Values for money: the technique of cost benefit analysis.
London: Gower Press, 1971. 237p.

6.389 GOLD, Bela
'Tracing gaps between expectations and results of
technological innovation: the case of iron and steel'. The
Journal of Industrial Economics, September, 1976. pp.1-28.

6.390 GOLD, Bela et al
Evaluating technological innovations: methods, expectations
and findings. Lexington, Mass.: Lexington Books, 1980.
358p.

6.391 HARVEY, Douglas C. and MENCHEN, Robert
The automobile, energy and the environment: a technology
assessment of advanced automotive propulsion systems.
Columbia, Maryland: Hittman Associates Inc., 1974. 159p.

162

6.392 HAUSTEIN, Heinz-Dieter and WEBER, Mathias
Selection and evaluation of innovation projects. Laxenburg:
I.I.A.S.A., 1980. 68p. (Working paper - 80-151).

6.393 KATZ, Milton
The function of tort liability in technological assessment.
Cambridge (Mass.): Harvard University, 1969. 587-662p.
(Harvard University Program on Technology and Society
Reprint no. 9)

6.394 KRANZBERG, Melvin
Historical aspects of technology assessment. Washington,
George Washington University 1969. Program of Policy Studies
in Science and Technology, Occasional paper, 4.

6.395 LAMSON, Robert W.
Framework of categories for science policy analysis and
technology assessment. Washington, Office of Planning and
Policy Studies, National Science Foundation, 1969. Mimeo.

6.396 MARX, Barbara Spence
Early experiences with the hazards of medical use of
x-rays: 1896-1906, a technology assessment case study.
Washington: George Washington University, 1968. (Program of
Policy Studies in Science and Technology, Staff Discussion
Paper, 205).

6.397 SURREY, A.J. and THOMAS, S.D.
Worldwide nuclear plant performance: lessons for technology
policy. SPRU Occasional Paper No.10. Brighton: SPRU, 1980.

6.398 THOMAS, S.D.
Worldwide nucleear plant performance revisited: an analysis
for the period 1978-1981. SPRU Occasional Paper Series No.
18. Brighton: SPRU, 1983.

6.d Other Aspects Of Management

6.399 ALBU, Austen
'Experiment in management'. Aspect, January 1964, pp.1-10.

6.400 ARNOLD, Erik
'New office technology - new problems'. The British Journal
of Administrative Management, Vol. 31, No. 7, October 1981,
pp.229-232.

6.401 BARNES, Louis B.

Organisational systems and engineering groups: a comparative study of two technical groups in industry. Boston: Harvard University, 1960.

6.402 BAUMOL, WILLIAM J.
Economic theory and operations analysis. 2nd ed. Englewood Cliffs, N.J., Prentice-Hall, 1965.

6.403 BILLSBORROW, Richard E. and PORTER, Richard C.
'The effects of tax exemption on investment by industrial firms in Columbia'. Weltwirtschaftliches Archiv: Review of World Economics, Vol.108, No.3, 1972, pp. 396-426.

6.404 BIRCHALL, D. and HAMAND, V.
Tomorrow's Office Today: managing technological change. Business Books Ltd: London, 1981.

6.405 BLAIR, Larry M.
Mechanisms for aiding worker adjustment to technological change. Vol.1. Conceptual Issues and Evidence. Final Report to the National Science Foundation. Salt Lake City: Human Resources Institute, University of Utah, 1974. Not continuously paginated.

6.406 BLOOD, Jerome W.
Utilizing R & D by-products. New York: American Mangement Association, 1967.

6.407 BLUME, Stuart S.
'Behavioural aspects of research management - a review'. Research Policy, Vol.3, No 1, p.40-76, April 1974.

6.408 BRIGHT, James Rieser
'Opportunity and threat in technological change'. Harvard Business Review, November/December 1963, 41(6), 76-86.

6.409 BRITISH SCIENTIFIC INSTRUMENTS RESEARCH ASSOCIATION
SIRA Conference on Innovation for Profit. Held at Eastbourne 6-8 May 1968. London, Hilger, 1968.

6.410 BRODIN, Bengt and VALDELIN, Jan
Market information and the product development process: a research project. Paper prepared for Seminar on New Product Management, European Institute for Advanced Studies in Management, Brussels, March 20-1, 1972. 10 leaves. Mimeo.

6.411 BROWN, Rosemary
'Training for the new technology'. Management Today, April 1981, p.104 and p.108.

6.412   BURACK, Elmer and MCNICHOLS, Thomas J.
Management and automation research project: final report.
Chicago, Illinois Institute of Technology, 1968.

6.413   BURNS, Tom and STALKER, G.
The Management of Innovation. 2nd Ed. London, Tavistock,
1961.

6.414   CHANDLER, Alfred D. and DAEMS, Herman (eds)
Managerial hierarchies: comparative perspectives on the
rise of the modern industrial enterprise.   Cambridge,
Mass./London: Harvard University Press, 1980. 237p.

6.415   CODDINGTON, Dean C. and others
Project for the analysis of technology transfer: annual
report, 1969. Denver, Industrial Economics Division, Denver
Research Institute, 1970. N70-25836.

6.416   COLLIER, D.W.
'An innovation system for the larger company'. Research
Management, 1970, 13(5), pp.341-349.

6.417   CONFEDERATION OF BRITISH INDUSTRY
Technology - putting it to work: opportunities, prospects,
problems. London: CBI, 1982. 47p.

6.418   COOK, F.W.
'Venture management as a new way to grow'. Innovation,
October 1971, p. 28-37.

6.419   COYLE, R.G.
'On the scope and purpose of industrial dynamics'.
International Journal of Systems Science, Vol.4, No. 3,
1973, p. 397-406.

6.420   DAFT, Richard L. and BECKER, Selwyn W.
Innovation in organizations: innovation adoption in school
organisation. New York, Oxford: Elsevier, 1978. 277p.

6.421   DAVIES, Duncan, BANFIELD, Tom and SHEAHAN, Ray
The humane technologist. London: O.U.P., 1976. 180p.

6.422   DOBROV, G.M.
'Technology as a form of organization', The International
Social Science Journal, Vol.XXXI, No.4, 1979, pp.585-605.

6.423   DROR, Israel
Strategic management in high-technology organizations.
Evanston, Illinois: Department of Industrial Engineering
and Management Sciences, The Technological Institute,

Northwestern University, 1981. 12 leaves. Mimeo. (Program of Research on the Management of Research and Development).

6.424 DUX, E.F.W.
'Personal problems in an R & D system'. Chemistry in Britain, September 1968, Vol.4(9), 393-396.

6.425 EDWARDS, G.A.B.
Readings in group technology: cellular systems.   Brighton: Machinery Publishing Co. Ltd., 1971. 140p.

6.426 EVANS, Alastair
What next at work? A new challenge for managers.   London: Institute of Personnel Management, 1979. 165p.

6.427 EYRING, H.B.
'Some sources of uncertainty and their consequences in engineering design projects'.   IEEE Transactions on Engineering Management, December 1966, Vol. EM-13.

6.428 FERRARI, Sergio
'Management for creativity'. European Business,   October 1968, No.19, p.p.38-44.

6.429 FINKELSTEIN, Stan N., SCOTT, Jeffrey R. and FRANKE, Ann
Diversity as a contributor to innovative performance by academic physicians. Cambridge, Massachusetts: Sloan School of Management, Massachusetts Institute of Technology. No date. 12 leaves. Mimeo.

6.430 FISHLOCK, David
The business of science: the risks and rewards of research and development. London: Associated Business, 1975. 169p.

6.431 FORD, David and RYAN, Chris
'Taking technology to market'.   Harvard Business Review, March-April, 1981, p.p.117-126.

6.432 FORES, Michael
The myth of technology and industrial science.   Berlin: International Institute of Management, 1979. 35p. (Discussion papers 79-49).

6.433 FORES, Michael and GLOVER, Ian (eds)
Manufacturing and management. London HMSO, 1978. 182p.

6.434 FORES, Michael and SORGE, Arndt
The decline of the management ethic. Berlin: International Institute of Management, 1979. 46p.

6.435 GALJAARD, J.H.
A technology based nation: an inquiry into industrial
organizing and robotizing in Japan. Delft: Inter-University
Institute of Management, Science and Technology Policy
Research Group, 1981. 155p. (Scitech Publication No. 3).

6.436 GANAPATHY, R.S.
'R & D, marketing, and social goals'. Economic and
Political Weekly, Vol.8, No.8, 24 February 1973. p.M7 - M12.

6.437 GARCIA-BOUZA, Jorge, SHAPERO, Albert and FERRARI,
       Achille
Technical entrepreneurship in Northern Italy. A preliminary
report. Milan: International Institute for the Management
of Technology, September, 1973. 65 leaves.

6.438 GERSTENFELD, Arthur and BERGER, Paul
'An analysis of utilization differences for scientific and
technical information'. Management Science, Vol.26, No.2,
February 1980, p.165-179.

6.439 GILLINGS, D.W.
'Process control economics: studies in the cost and
profitability of automatic process control'. Instrument
Practice, May-December 1964, 45pp.

6.440 GOLD, Bela
'Productivity analysis and system coherence'. Operational
Research Quarterly, Vol. 16(3), 1965, pp.287-307.

6.441 GOLD, Bela
'New perspectives on cost theory and empirical findings'.
Journal of Industrial Economics, Vol. 15, April 1966,
pp.164-198.

6.442 GOLD, Bela
'Productivity analysis: some new analytical and empirical
perspectives'. Business Economics, Vol. 9, No. 3, May 1974,
P. 64-72.

6.443 GOLD, Bela
Technological change: economics, management and environment.
Oxford: Pergamon Press, 1975. 175p.

6.444 GOLD, Bela.
'Values and research: the framework of decision for major
technological innovation'. In: Values and the Future,
edited by Kurt Baier and Nicholas Rescher. New York: Free
Press, 1969.

6.445 GOODE, Harry H.
'Intracompany systems management'. IRE Transactions on Engineering Management, Vol.EM-7(1), 1960. pp.14-19.

6.446 GORDON, Gerald and MORSE, Edward
'Creative potential and organizational structure'. Academy of Management Journal, Spring 1969. 49p.

6.447 GOUGH, J.W.
The rise of the entrepreneur. London, Datsford, 1969.

6.448 GREEN, Diana
Managing industrial change? French policies to promote industrial adjustment. Prepared for the Department of Industry. London: HMSO, 1981. 81p.

6.449 GRUBER, William H. and MARQUIS, Donald G. (eds)
Factors in the transfer of technology. Cambridge, Mass., MIT Press, 1969.

6.450 HARMAN, Alvin J.
Innovation management: research protocol for international collaboration. Draft paper. Laxenburg: IIASA, 1980. 46p. Mimeo.

6.451 HARVEY, Edward Burns
Structure and process in industrial organizations. Ph.D. thesis, Princeton University, 1967.

6.452 HETZLER, Stanley
Applied measures for promoting technological growth. London: Routledge and Kegan Paul, 1973. 337p.

6.453 HOGARTH, Robin
'Problems in the marketing of innovations'. European Business, January 1969, No. 20, 17-22.

6.454 HOLT, Knud
'Management of technological innovation'. Management International Review, 1970.

6.455 HU, Y.S.
Corporate planning in Europe: the market threshold. Manchester: Manchester Business School, 1972. 20 leaves.

6.456 HUGHES, Thomas D.
Transfer and style: an historical account. Paper presented at the International Conference on Technology Transfer Control Systems: issues, perspectives, implications. Philadelphia, 9-10 February, 1979. No imprint, 1979. 20p.

Mimeo.

6.457   INHABER, H. and LIPSETT, M.S.
'Gaps in "gaps in technology" and other innovation
inventories'. Scientometrics, Vol.1, No.5-6, August 1979,
pp.405-417.

6.458   IWANOWSKA, Anna
Two approaches to the study of innovation problems. Berlin:
International Institute of Management, 1979. 19 leaves.
(IIM/dp79-51).

6.459   JARMAIN, W. Edwin (ed)
Problems in industrial dynamics. Cambridge, Mass.:   M.I.T.
Press, 1963. 124p.

6.460   JOHNSTON, Ron and GUMMETT, Philip (eds)
Directing technology: policies for promotion and control.
London: Croom Helm, 1979. 271p.

6.461   JONES, P.M.S.
'Market research in the novel product field'. IMRA Journal,
Vol. 7, No. 1, p. 32-47, February 1971.

6.462   KELNER, H.W.
'Assessment of markets for entirely new products'. IMRA
Journal, November 1970, Vol. 6, No. 4, p. 169-180.

6.463   KOBAYASHI, Tatsuya
The mode of technology transfer: some features induced from
Japanese experiences. Chukyo University, no date. 14 leaves.
Mimeo.

6.464   LAMBRIGHT, W. Henry
Governing science and technology. London: O.U.P.,   1976.
218p.

6.465   LAMSAC
Managing technology in the '80's.   2   volumes.   London:
LAMSAC, no date. 16p. + 161p.

6.466   LAWRENCE, Peter
Managers and management in West Germany. London: Croom
Helm, 1980. 202p.

6.467   LEARSON, T.Vincent
'The management of change'. Columbia Journal of World
Business, January-February 1968, 59-64.

6.468   LESTER, Tom

'The micro future.....now.' Management Today, January 1980. pp.42-49.

6.469 LITTLER, D.A. (ed)
The management of industrial innovation. Proceedings of a Seminar held at Liverpool University on 7th December, 1977. 97 leaves. Mimeo.

6.470 LODAHL, Thomas M. and MEYER, N. Dean
'Six pathways to office automation'. Administrative Management (USA), March 1980, pp.32-33, 74, 78, 80.

6.471 LORSCH, Jay W.
Product innovation and organization. London: Collier-MacMillan, 1965. 184p.

6.472 LORSCH, Jay W. and LAWRENCE, Paul R.
'Organizing for product innovation'. Harvard Business Review, January/February 1965, Vol. 34(1), 109-122.

6.473 MACRURY, King
'Command posts for diversification strategy'. Financial Executive, February 1968, 76-80.

6.474 MAIDIQUE, Modesto A.
'Entrepreneurs, champions and technological innovation'. Sloan Management Review, Vol.21, No.2, Winter 1980, pp.59-76.

6.475 MANT, Alistair
The rise and fall of the British manager. London: Mac-Millan, 1977. 142p.

6.476 MCNEIL, Ian
'It's time Britain put robots in their place'. Works Management, July 1979, pp.54-6, 59-60.

6.477 MENCHER, Alan
'The true technology gap', Management Today, October 1967,p. 76-69, 146.

6.478 MENCHER, Alan
Management and technology-volume 1: an Anglo-American exchange of views. Frimley: Infolink, 1972. (SPF Special Publications series)

6.479 MILLIKEN, J. Gordon and MORRISON, Edward J. & CRISTIANO, Carole R.
Aerospace contributions to management: selected cases illustrating significant management techniques and computer

software transferred from U.S. air & space programs. Denver:
Denver Research Institute, 1975. 262 leaves.

6.480 MOTTLEY, Charles M
'Managing innovation for growth', Stanford Research
Institute Journal, 1961, Vol. 5, 58-66.

6.481 MUMFORD, Enid and HENSHALL, Don
A participative approach to computer systems design: a case
study of the introduction of a new computer system. London:
Association Business Press, 1979. 187p.

6.482 MUMFORD, Enid et al
'The human problems of computer introduction'. Management
Decision, Vol.10, Spring 1972, pp.6-17.

6.483 NORTH, Jeremy
'Reducing the risk of innovation'. Business Administration,
March 1970, 33-35.

6.484 PACIFICO, C.
'Is it worth the risk?' Chemical Engineering Progress, May
1964, 60, 19-21.

6.485 PARKER, R.C.
Guidelines for product innovation. London: British Institute of Management Foundation, 1980. 56p.

6.486 POLYTECHNIC SCHOOL OF MANAGEMENT STUDIES
Report on the proceedings of a seminar on the economics of
research and development at the Polytechnic School of
Management Studies, January 1967. pp.150-154, bibliography
on licensing, patents, trade marks, research, investment and
the relevant law.

6.487 RANDALL, Geoffrey
Managing new products. London:BIM, 1980. 24P. (Management
Survey Report No.47).

6.488 ROBERTSON, Andrew
'Innovation management'. Management Decision Monograph,
12(6), pp.329-372.

6.489 ROBERTSON, Andrew
The management of industrial innovation: some notes on the
success and failure of innovation. London, Industrial
Educational and Research Foundation, 1969. Occasional Paper,
7.

6.490 ROBERTSON, Andrew

'The marketing factor in successful industrial innovation'. Industrial Marketig Management, Vol.2, 1973, pp.369-374.

6.491 ROSENBLOOM, Richard S.
Technology transfer - process and policy: an analysis of the utilization of technological by-products of military and space R & D. Washington, National Planning Association, 1965. Special report No. 62.

6.492 ROSENBLOOM, Richard S. and ABERNATHY, William J.
The climate for innovation in industry: the role of management attitudes and practices in consumer electronics. Boston, Mass.: Harvard University, Graduate School of Business Administration, 1980. 36 leaves. Mimeo. (Working Paper HBS 80-65 Rev. 1/82).

6.493 ROSS, J.A.
The implementation and servicing of a licence agreement. Mimeo. Paper presented at the Management of International Licensing Conference, London 5-6 June, 1969.

6.494 ROWLAND, Tom
'Technology's threatening promise'. Management Today, March 1979, pp.94-7.

6.495 SADLER, P.J. and BARRY, B.A.
Organisational development: case studies in the printing industry. London, Longmans, 1969.

6.496 SAHAL, Devendra
'The determinants of best-practice technology'. In: R & D Management Vol.11, No.1, 1981, pp.25-32.

6.497 SAHAL, Devendra
Structure and self-organisation. New York: New York University, Graduate School of Business Administration, 1982. 27 leaves. Mimeo.

6.498 SAHAL, Devendra (ed)
Research, development, and technological innovation: recent perspectives on management. Lexington, Mass.: Lexington Books, 1980. 274p.

6.499 SCHNEE, Jerome F.
'Development cost: determinants and overruns'. Journal of Business Vol.45, 1972, pp.347-374.

6.500 SENKER, Peter J.
'So who needs designers anyway?' Design, No.369, September 1979, pp.69-70.

6.501 SKINNER, R.N.
<u>Launching new products in competitive markets.</u>   London: Cassell/Associated Business Programmes, 1972. 184p.

6.502 SOLO, Robert A.
<u>Organizing science for technology transfer in economic development.</u> Michigan: Michigan State University Press, 1975. 224p.

6.503 SOUDER, William E.
<u>An R & D planning and control system: a case history.</u> Paper presented in part at the 37th National Operations Research Society Meeting, Washington D.C., April 20, 1970. 36 leaves. (Program of Research on the Management of Research and Development - Project 10). Mimeo.

6.504 SOUDER, William E.
<u>Comparative analysis of R & D investment models.</u> Evanston, Illinois: Northwestern University, 1971. 12 leaves. (Program of Research on the Management of Research and Development - Project 10). Mimeo.

6.505 SOUDER, William E., MAHER, Michael and RUBENSTEIN, A.H.
<u>R & D project selection: results from two experiments.</u> Evanston, Illinois: Northwestern University, 1971. 18 leaves. (Program of Research on the Management of Research and Development - Project 10)·. Mimeo.

6.506 STEAD, Humphrey
'The costs of technological innovation'. <u>Research Policy</u>, Vol.5, No.1, January 1976, pp.2-9.

6.507 STEPHANOW, S.E.
<u>Management: technology, innovation and engineering.</u> Malibu, California: Daniel Spencer Publishers, 1980. 391p.

6.508 SWORDS-ISHERWOOD, Nuala
<u>Management requirements for successful change.</u> Paper prepared for Manpower Society Conference "The challenge of change", September 1979. Brighton: SPRU, August 1979. 24 leaves. Mimeo.

6.509 TEIRE, John and VARNEY, David
'Social values and technological change: the shift working dilemma'. <u>Management Decision</u>, Vol.9, No.2, Summer 1971, pp.179-187.

6.510 THOMAS, Michael J.
'Making R & D pay: the role of market research'. <u>Chemistry</u>

and Industry, 18 August 1973, pp.766-8.

6.511 THOMPSON, Victor A.
'Bureaucracy   and   innovation'.   Administrative Science
Quarterly 10(1), 1965, pp.1-20.

6.512 TOMKINS, Raymond
Educating the engineer for innovative and entrepreneurial
activity. A study carried out for the Commission of the
European Communities by the Societe Europeene pour la
Formation des Ingenieurs (SEFI). London: Imperial College
of Science and Technology, Department of Management Science,
1979. 146p. Mimeo.

6.513 TWISS, Brian C.
Managing technological innovation. London: Longman, 1974.
235p.

6.514 UNIVERSITY OF BRADFORD - Management Centre
National Conference on the Management of Technological
Innovation, 12 & 13 March 1969 - a conference manual.
Bradford: The University, 1969. 112 leaves.

6.515 UTTERBACK, James M.
Management of technology. Cambridge, Mass.: Centre for
Policy Alternatives, Massachusetts Institute of Technology,
1978. 26p. Mimeo.

6.516 VEDIN, Bengt-Arne
Creativity management in media industry applied to
technology. Lund, Sweden: Institute for Management of
Innovation and Technology, 1980. 97p. Mimeo.

6.517 VON HIPPEL, Eric
'Has a customer already developed your next product?' Sloan
Management Review, Vol. 18, No. 2, Winter 1977, p. 63-74.

6.518 VON HIPPEL, Eric
'Successful and failing internal corporate ventures: an
empirical analysis'. Industrial Marketing Management, Vol.
6, 1977, p. 163-174.

6.519 VON HIPPEL, Eric
Increasing innovators' returns from innovations. Cambridge,
Mass.:   Alfred P. Sloan School of Management, Massachusetts
Institute of Technology, 1981. 22p. Mimeo. (Working Paper
No. 1192-81).

6.520 WHITE, George R.
Management criteria for technological innovation.

Transcription from "The management of innovation: an MIT Sloan School Symposium", 9 December 1976, New York. No imprint. 1976. 44 leaves. Mimeo.

6.521 WHITE-HUNT, K. and MAKOWER, M. (eds)
The encouragement of technological change. Proceedings of the Leverhulme Symposium, proceedings, papers and discussion held on 14 November 1980. Stirling: Technological Economics Research Unit, Department of Management Science and Technology Studies, University of Stirling, 1980. 77p. Mimeo. (Discussion Paper No.27).

6.522 WILSON, A.C.B.
'Planning in the medium-sized'. Management Decision, 1968, pp.28-32.

6.523 WILSON, Ira G. and WILSON, Marthann E.
Management, innovation and system design. Princeton, London: Auerbach, 1971. 175p.

CHAPTER 7. <u>Technical Change, Innovation And Government</u>

7.a   The Effect Of Government Policies On Innovation

7.001   ABERNATHY, William J. and CHAKRAVARTHY, Balaji S.
'Government intervention and innovation in industry: a
policy framework'. <u>Sloan Management Review</u>, Spring 1979,
pp.3-18.

7.002   BANNOCK, Graham and DOREN, Alan
<u>Government barriers to innovation by small firms in the U.K.</u>
Prepared for the Commission of the European Communities
Directorate General Scientific and Technical Information and
Information Management. London: Economists Advisory Group
Ltd., 1979. 90p.

7.003   BAR-ZAKAY, Samuel N.
'Policy making and technology transfer: the need for
national thinking laboratories'. <u>Policy Sciences</u>, Vol. 2,
1971, p.213-227.

7.004   BLACK, Guy
<u>Some etfects of federal procurement of research and</u>
<u>development on industry.</u> Washington, D.C.: Program of
Policy Studies in Science and Technology, George Washington
University, 1968.

7.005   BLACK, Guy
<u>The etfect of government funding on commercial R & D.</u>
Washington DC: Program of Policy Studies in Science and
Technology, George Washington University. 1970. 16p.

7.006   BOUCHER, Wayne I. and others
<u>Federal incentives for innovation. The impact of EPA</u>
<u>administrative practice on the innovation process in U.S.</u>
<u>companies.</u> A case study of regulatory barriers to
innovation. Denver: Denver Research Institute, 1976.
(Report R 75-05).

7.007   BRAUNLING, G., MULLER, P. and STROEIMANN, K.A.
<u>Towards an assessment of government measures to promote</u>
<u>technical change in industry - a framework of criteria.</u>

First results of the Central Project of the Six Countries Programme on "Government Policies towards Technological Innovation in Industry". Karlsruhe: ISI, 1976. 57 leaves. Mimeo.

7.008 BROZEN, Yale
'The role of government in research and development'. American Behavioral Scientist, 1962, 6(4), 22-26.

7.009 BRUM, Harold J. and HEMPHILL, John H.
The role of government in the allocation of resources to technologial innovation. Final report submitted to: Office of National R & D Assessment, National Science Foundation 1975 - 1976. Vol.I Final report. Vol.II Survey of the literature on policy guidelines and agenda for future research. Vol. III Abstracts, annotated bibliography.

7.010 BURSTALL, M.L., DUNNING, M. and LAKE, A.
Multinational enterprises, governments and technology: a study of the impact of multinational enterprises on national scientific and technical capacities in the pharmaceutical industry. Paris: OECD, 1981. 252p.

7.011 CAIRNCROSS, Sir Alexander
Government and innovation. Presidential address delivered to the Swansea meeting of the British Association, September 1, 1971. 15p.

7.012 CANADA - Science Council of Canada
Governments and innovation. By Andrew H. Wilson. Ottawa: Information Canada, 1973. 275p. (Special Study no. 26).

7.013 CANADA - Science Council of Canada
The role and function of government laboratories and the transfer of technology to the manufacturing sector. By Arthur J. Cordell and James Gilmour. Ottawa: Information Canada, 1976. 397p. (Background study no.35).

7.014 CARMICHAEL, Jeffrey
'The erfects of mission-oriented public R & D spending on private industry'. The Journal of Finance, Vol.XXXVI, No.3, June 1981, p.617-627.

7.015 CARTER, Charles (ed)
Industrial policy and innovation. London: Heinmann, 1981. 241p. (Joint Studies in Public Policy 3).

7.016 CARTER, Charles F. and WILLIAMS, B.R.
'Government scientific policy and the growth of the British economy'. Minerva, 3(1), 1964, pp.114-125.

7.017 CHAPMAN, Gordon
Program budgeting as a tool for scientific and technologial policy and planning. Mimeo. Washington, Department of Scientific Affairs, Organisation of American States, 1972.

7.018 CHRISTIANSEN, Gregory B. and HAVEMAN, Robert H.
'Public regulations and the slowdown in productivity growth'. American Economic Review, Vol.71, No.2, 1981. p.320-325.

7.019 CLARKE, Sir Richard
Aims and means of a European policy on technological development. Paper presented at the Conference on Industry and Society in the European Community. Venice, 1972. 28p. (Report no.8.)

7.020 CORDES, Joseph J.
The impact of tax and financial regulatory policies on industrial innovation. Washington DC: National Academy of Sciences, 1980. 34p. (Technology, Trade and the U.S. Economy).

7.021 DANHOF, Clarence H.
Government contracting and technological change. Washington: Brookings Institute, 1968. 472p.

7.022 DAVEY, D.G.
'The concern over drug safety and its implications for the industry's innovation and growth'. Chemistry and Industry, 2 June 1973, 508-9.

7.023 DENVER RESEARCH INSTITUTE
Ranking priorities for experimentation with federal incentives to innovation - a methodology. Denver: Denver Research Institute, 1975. 54p.

7.024 DOCTORS, Samuel I.
The role of Federal agencies in technology transfer. Cambridge, Mass.: MIT Press, 1969.

7.025 DUCHENE, Francois and SHEPHERD, Geoffrey
Industrial adjustment and government intervention in Western Europe. Brighton: SERC, University of Sussex, October 1980. 61p. Mimeo.

7.026 EADS, George
'U.S. government support for civilian technology, economic theory versus political practice'. Research Policy, Vol.3, No.1, April 1974. pp.2-16.

7.027   EADS, George and NELSON, R.
Governmental support ot advanced civilian technology -
power reactors and the supersonic transport.   New   Haven:
Economic Growth Center, Yale University, 1970. 27 leaves.
(Economic Growth Centre Discussion Paper No.102).

7.028   ECKERT, Theodore J.
The transfer ot U.S. technology to other countries: an
analysis of export control policy and some recommendations.
Princeton: Princeton University, Woodrow Wilson School of
Public   and   International   Affairs, 1981. 60p. Mimeo.
(Research Monograph No. 47).

7.029   EIRMA
Industry - government relations in R & D.   Paris: EIRMA,
1975. 94p. (EIRMA Working Group Reports no. 15.).

7.030   ELIASSON, Gunnar and YSANDER, Bengt-Christer
Picking winners or bailing out losers?   Stockholm: IUI,
1981. 70 leaves. Mimeo.

7.031   ENCEL, Sol and RONAYNE, Jarlath (eds)
Science, technology and public policy: an international
perspective. Rushcutters Bay, N.S.W., Oxford: Pergamon
Press, 1979. 159p.

7.032   EUROPEAN TRADE UNION INSTITUTE
Industrial policy in Western Europe. Brussels: ETUI, 1981.
4 leaves. Mimeo.

7.033   EUROPEAN TRADE UNION INSTITUTE
European industrial policy. Report to EFTA on economic
policies which affect industrial structure and trade.
Bruxelles: ETUI, 1981. 291p.

7.034   EVANS, M. David
A general equilibrium analysis of protection: the effects
of protection in Australia.   Amsterdam,   London: North-
Holland, 1972. 216p.

7.035   FAIRLAMB, David
'Public policy issues and the technological revolution'.
The Banker, Vol.131, No. 622, April 1981, p.113-115.

7.036   FOURNIER, Rene-Paul
Canadian experience in government measures to promote
technological innovation. 17 leaves. SPRU. 1974. Mimeo.

7.037   FREEMAN, C.
Government policies for industrial innovation. J.D. Bernal

Memorial Lecture, Birkbeck College, 23 May 1978. 16 leaves.
Mimeo.

7.038 GERSTENFELD, Arthur and BRAINARD, Robert (eds)
<u>Technological innovation: government/industry cooperation.</u>
New York, Chichester: John Wiley & Sons, 1979. 277p.

7.039 GILPIN, Robert
'Technological strategies and national purpose'. <u>Science,</u>
169(3944), 31 July 1970, pp.441-448.

7.040 GINSBURG, Douglas H.
<u>Antitrust, uncertainty, and technological innovation.</u>
Washington DC: National Academy of Sciences, 1980. 51p.

7.041 GORDON, John Elliott
<u>The effects of government sponsored R & D on the management</u>
<u>of private R & D.</u> Microfilm. Thesis, M.I.T., 1966.

7.042 GRABOWSKI, Henry G.
<u>Drug regulation and innovation: empirical evidence and</u>
<u>policy options.</u> Washington, D.C.: American Enterprise
Institute for Public Policy Research, 1976. 82p.

7.043 GRANGER, John V.
<u>Technology and international relations.</u> San Francisco, W.H.
Freeman & Co., 1979. 202p.

7.044 GREEN, Diana
<u>Managing industrial change? French policies to promote</u>
<u>industrial adjustment.</u> Prepared for the Department of
Industry. London: HMSO, 1981. 81p.

7.045 GREEN, Harold P.
'Nuclear technology and the fabric of government'. George
Washington University, 1965. Program of Policy Studies in
Science and Technology, Paper 7. Reprinted from <u>George</u>
<u>Washington Law Review</u> No.1, October 1964.

7.046 GREENBERG, Daniel S. (ed)
<u>Science and Government Report International Almanac,</u>
<u>1978-1979.</u> Washington, D.C.: Science and Government Inc.,
1979. 368p.

7.047 GUMMETT, Philip and GIBBONS, Michael
'Government research for industry: recent British
developments'. <u>Research Policy</u>, Vol.7, 1978. p. 268-290.

7.048 HAGLAND, Howard H. and SCHLIE, Theodore W.
<u>Federal incentives for innovation: preliminary findings in</u>

the development of an experiment to realistically explore
and promote the transfer of technology from Federal Labs.
Denver: Denver Research Institute, 1975. 93p. (Report R
75-02).

7.049 HELLER, Terry M. and MATTHEWS, William E.
Federal incentives for innovation. Summary of background
studies relating to the stimulation of innovation. Denver:
Denver Research Institute, 1976. 97p. (Report R 75-03).

7.050 HOLMAN, Mary A.
'Patent policies of other governments'. Idea, Spring 1964,
Vol. 7, 94ff.

7.051 HOLMAN, Mary A.
'Government research and development inventions: a new
resource'. Land Economics, Vol. 41, August 1965, pp.231-238.
N65-88511.

7.052 INSTITUT FUR SYSTEMTECKNIK UND INNOVATIONSFORSCHUNG
Impact analysis of the Federal subsidies for funding
research and development personnel in small and medium-sized
enterprises - interim report. By Gisela Gielow et al.
Karlsruhe: Institute fur Systemtechnik und
Innovationsforschung, 1981. 20p.

7.053 JOSEPH, R.A.
The industrial research and development grants scheme - a
study in Australian government innovation. School of
Science, Griffith University, Science, Technology and
Society Project, 1977. 66p. Mimeo.

7.054 JOSEPH, R.A.
A comparative study of Federal government incentive
programmes for industrial research and development in
Australia and Canada. Australia: Griffith University, 1978.
89p. Mimeo.

7.055 KAYSEN, Carl
'Basic research and national goals: a report to the
Committee on Science and Astronautics'. Minerva, 1965, Vol.
4(2), 254-272.

7.056 KECK, O.
'The West German fast breeder programme: a case study in
governmental decision making'. Energy Policy, Vol.8, No.4,
December 1980.

7.057 KING, Alexander
'Science policy, economic growth and the quality of life'.

Science Policy News, 2(1), July 1970, pp.1-6.

7.058 KNIGHT, Kenneth E., KOZMETSKY, George and BACA, Helen R.
Industry views of the role of the Federal government in industrial innovation. Austin: Graduate School of Business, University of Texas, 1976. 127p.

7.059 KRISTENSEN, Peer Hull and STANKIEWICZ, Rikard (eds)
Technology policy and industrial development in Scandinavia. Proceedings from a workshop held in Copenhagen, Denmark, May 20-21, 1981. Lund: Research Policy Institute & Roskilde: Institute of Economics and Planning, 1982. 218p.

7.060 KUEHN, Thomas J.
'Technological innovation and public policy'. *The Trend in Engineering at the University of Washington*, Vol. 26, No. 2, April 1974, p. 29-33.

7.061 KUEHN, Thomas J. and PORTER, Alan L. (eds)
Science, technology and national policy. Ithaca and London: Cornell University Press, 1981. 530p.

7.062 LAKI, Mihaly
The function of the state in the introduction of new products and new technologies. Budapest: Institute for Economic & Market Research, 1977. 31 leaves.

7.063 LITTLE, Blair
The role of government in assisting new product development. London, Ontario: University of Western Ontario, 1974. 8 leaves and appendices. Mimeo. (Working paper series No. 114).

7.064 LIU, Laurence and KAMATH, Dinesh
Regulation and technology choice in telecommunications - final report. San Rafael, Ca.: Systems Applications, 1977. 130p + Appendices.

7.065 MACDONALD, A.S.
'Exchange rates for national expenditure on research and development'. *The Economic Journal*, June 1973, p. 477-494.

7.066 MADDOCK, I.
Stimulating technological innovation in industry: the role of the Ministry of Technology. Paper to be presented to the Institution of Civil Engineers, London, 18th June 1968. London: The Institution of Mechanical Engineers, 1968. 11p.

7.067 MANSFIELD, Edwin

'Some recent economic studies bearing on public policy toward civilian technology'. Managerial and Decision Economics, Vol. 1, No. 4, 1980 p. 167-171.

7.068 MASSACHUSETTS INSTITUTE OF TECHNOLOGY - Center for
      Policy Alternatives
National support for science and technology: an examination of foreign experience. 2 volumes. Cambridge Mass.: Center for Policy Alternatives, 1975. Not paginated.

7.069 MAUNDER, Peter (ed)
Government intervention in the developed economy. London: Croom Helm, 1979. 226p.

7.070 MISRA, V. K.
A study of Japanese policy instruments for promoting research and development with special reference to electronics. New Delhi: Electronics Commission Information Planning & Analysis Group, 1976. 42p. (Area survey report 33) Mimeo.

7.071 MOONMAN, Eric
British computers and industrial innovation: the implications of the Parliamentary Select Committee. London: Allen & Unwin, 1971. 126p.

7.072 MOWERY, David C. and ROSENBERG, Nathan
Government policy and innovation in the commercial aircraft industry, 1925-1975. Preliminary draft. Stanford, California: Stanford University, 1980. 73 leaves. Mimeo.

7.073 MUSKIE, Edmund S.
'The role of Congress in promoting and controlling technological advance'. George Washington Law Review, 36(5), July 1968, pp.1138-1149.

7.074 MYERS, Sumner and SWEEZY, Eldon
Federal incentives for innovation. Why innovations falter and fail: a study of 200 cases. Denver: Denver Research Institute, 1976. 77p. (Report R 75-04).

7.075 McCULLOCH, D.G.
Technology and innovation in New Zealand manufacturing industry. 'The role of government'. Resource Paper 2, Lower Hutt, New Zealand: Physics and Engineering Laboratory, DSIR, 1980. 66p. Mimeo. (PEL Report No. 686)

7.076 NARASIMHAN, R.
Technological change and government entrepreneurship. Bombay: National Centre for Software Development and

Computing Techniques, Tate Institute of Fundamental Research, 1978. 28 leaves. Mimeo.

7.077 NELSON, Richard R.
Balancing market failure and governmental inadequacy: the case of policy toward industrial R & D. New Haven: Yale University Institution for Social and Policy Studies. No date. 29 leaves and appendices. Mimeo. (Working Paper No. 840).

7.078 NELSON, Richard R.
The "technology gap" and national science policy. Economic Growth Centre, Yale University, 1970. Mimeo.

7.079 NELSON, Richard R.
Government stimulus of technological progress - lessons from American History. New Haven: Yale University Institution for Social and Policy Studies, 1981. 51 Leaves. Mimeo.

7.080 NELSON, Richard R.
Balancing market failure and governmental inadequacy: the case of policy toward industrial R & D. New Haven, Conn.: Institution for Social and Policy Studies. No date. 29 leaves & appendices. Mimeo. (Working Paper No. 840)

7.081 PALDA, Kristian S.
The Science Council's weakest link: a critique of the Science Council's technocratic industrial strategy for Canada. Vancouver: The Fraser Institute, 1979. 60p. Pbk.

7.082 PAVITT, K.
'Governmental support for industrial research and development in France: theory and practice'. Minerva, Vol.XIV, No.3, Autumn 1976, pp.330-335.

7.083 PAVITT, Keith
'Government policies towards innovation: a review of empirical findings'. Omega, Vol.4, No.5, 1976, pp.539-558.

7.084 PAVITT, Keith
'Governmental support for industrial innovation: the Western European experience' in: R. Johnston and P. Gummett (eds)
Directing technology: policies for promotion and control. London: Croom Helm, 1979. Chapter 1.

7.085 PAVITT, Keith and WALKER, William
'Government policies towards industrial innovation'. Research Policy, Vol. 5, No. 1, January 1976. pp.11-97.

7.086   QUINN, James Brian
'National strategy in science and technology'. Science
Journal, April 1969, 5(4), 77-81.

7.087   ROBBINS, Martin D. and MILLIKEN, J. Gordon
'Government policies for technological innovation: criteria
for an experimental approach'. Research Policy, Vol.6, No.3,
July 1977, pp.214-240.

7.088   ROTHWELL, Roy
'The impact of regulation on innovation: some U.S. data'.
Technological Forecasting and Social Change, Vol.17, No.1,
May 1980, pp.7-34.

7.089   ROTHWELL, Roy
Industrial innovation and government regulation. Report to
the Six Countries Programme on Aspects of Government Policy
towards Innovation in Industry and the National Science
Foundation. Brighton: SPRU, March 1980. Individual
pagination.

7.090   ROTHWELL, Roy
'Pointers to government policies for technical innovation'.
Futures Vol.13, No.3, June 1981, pp.171-183.

7.091   ROTHWELL, Roy
'Some indirect impacts of government regulation on
industrial innovation in the United States'. Technological
Forecasting and Social Change, Vol.19, No.1, February 1981,
pp.57-80.

7.092   ROTHWELL, Roy
'Government innovation policy: some past problems and
recent trends', Technological Forecasting and Social Change,
Vol.22, No.1, September 1982, pp.3-30.

7.093   ROTHWELL, Roy and WALSH, V.
Regulation and innovation in the chemical industry.   Paper
prepared for OECD Workshop, Paris, 20th and 21st September,
1979. Brighton: SPRU, September 1979. Not paginated. Mimeo.

7.094   ROTHWELL, Roy and ZEGVELD, Walter
Small and medium sized manufacturing firms: their role and
problems in innovation government policy in Europe, the USA,
Canada, Japan and Israel. Report prepared for the Six
Countries Programme on Government Policies towards
technological innovation and industry. Vol.1. June 1978.
171p. Mimeo.

7.095   ROTHWELL, Roy and ZEGVELD, Walter

Industrial innovation and public policy: preparing for the 1980's and the 1990's. London: Francis Pinter, 1981. 251p.

7.096 ROTHWELL, Roy et al
'The identification of priority areas for research in the external aids field of biomedical engineering'. International Journal of Rehabilitation Research,1979, 2(4), 539–540.

7.097 RUBENSTEIN, Albert H. et al
'Management perceptions of government incentives to technological innovation in England, France, W. Germany and Japan'. Research Policy Vol.6, No.4, October 1977, pp.324–357.

7.098 RUBENSTEIN, C.L.
'Changes in Australian science and technology policies: from ends to means'. Australian Journal of Public Administration, Vol.37, No.3, September 1978. pp.232–256.

7.099 S.R.I. INTERNATIONAL
Management of Federal R & D for commercialization – executive summary. By Norman B. McEachron et al. Prepared for the Experimental Technology Incentives Program, Department of Commerce. Menlo Park, California: SRI International, 1978. 14 leaves.

7.100 SALOMON, Jean-Jaques
'Technical change and economic policy'. In: The OECD Observer, No.104, May 1980. pp.16–22.

7.101 SCHENK, W.
Micro-electronics in Austrian manufacturing: implications for a government policy promoting indigenous innovation. Paper submitted for IIASA Workshop on Innovation Management, 22–25 June, 1981. Vienna. Brighton: SPRU, June 1981. 11 leaves. Mimeo.

7.102 SCHWARZ, Michiel
'Government responsibility for technology: microelectronics policies in Western Europe'. In: Microprocessing and Microprogramming, Vol.8, 1981, pp.67–82.

7.103 SCHWEITZER, G.E. and PINSKY, P.
Contributions to industrial development of science and technology institutions in Colombia and opportunities for bilateral cooperation. Ithaca, New York: Cornell University, Program on Science, Technology and Society, 1979. 46p. Mimeo.

7.104 SIX COUNTRIES PROGRAMME
The current international economic climate and policies for
technical innovation. Report prepared for the Six Countries
Programme on Government Policies towards technological
innovation in industry, by Science Policy Research Unit,
University of Sussex in collaboration with Staffgroup
Strategic Surveys TNO, November 1977. 53 leaves. Mimeo.

7.105 SIX COUNTRIES PROGRAMME ON ASPECTS OF GOVERNMENT
       POLICIES TOWARDS TECHNOLOGICAL INNOVATION
Government direct financial assistance to industry:
programmes, experiences and trends. Executive   Summary
Report, by J. Knox. Delft, Netherlands: Staffgroup
Strategic Surveys TNO, February 1977. 33 leaves. Mimeo.

7.106 SIX COUNTRIES PROGRAMME ON ASPECTS OF GOVERNMENT
       POLICIES TOWARDS TECHNOLOGICAL INNOVATION IN
       INDUSTRY
Government procurement policies and industrial innovation.
Report prepared for the Six Countries Programme on
Government Policies towards Technological Innovation in
Industry, by W. Overmeer and F. Prakke. Delft, Netherlands:
Staffgroup Strategic Surveys TNO: December 1978. 82p.
Mimeo.

7.107 SIX COUNTRIES PROGRAMME ON ASPECTS OF GOVERNMENT
       POLICIES TOWARDS TECHNOLOGICAL INNOVATION IN
       INDUSTRY
Papers presented at a Workshop on the relationship between
Technical Development and Employment. Paris, 1978. Delft:
Six Countries Programme Secretariat, 1978. 368 leaves.
Mimeo.

7.108 SIX COUNTRIES PROGRAMME ON ASPECTS OF GOVERNMENT
       POLICIES TOWARDS TECHNOLOGICAL INNOVATION IN
       INDUSTRY
Workshop on Government (public) demand as an instrument to
stimulate innovation in industry. Dublin, June 6 & 7, 1978.
Various papers.

7.109 SIX COUNTRIES PROGRAMME ON ASPECTS OF GOVERNMENT
       POLICIES TOWARDS TECHNOLOGICAL INNOVATION IN
       INDUSTRY
Trends in collective industrial research. Papers presented
at the Workshop on Trends in Collective Industrial Research,
London, November 26-27, 1979. Delft, The Netherlands;
Secretariat Six Countries Programme, 1979. 184p. Mimeo.

7.110 SOLTYSINSKI, Stanislaw J.
The impact of new transnational technology transfer control

schemes upon the international patent system: a European
perspective. Paper prepared for the International Conference
on   Technology   Transfer   Control   Systems: issues,
perspectives, implications, Philadelphia, February 9-10,
1979. No imprint. 50p. Mimeo.

7.111 SUTMEIER, Richard P.
Research, innovation and science policy for development. New
York: Hamilton College, Clinton, 1972. 44p.

7.112 TABB, William K.
'Government regulations: two sides to the story'.
Challenge, November-December 1980, pp.40-48.

7.113 TEUBAL, Morris and STEINMUELLER, Edward
Government policy, innovation and economic growth: lessons
from a study of satellite communications.   (No   imprint).
1979. 68p.

7.114 TEUBAL, Morris and STEINMUELLER, Edward
Government policy and the introduction of a major new
technology: elements of a framework of analysis. Jerusalem:
The Morris Falk Institute for Economic Research in Israel,
1980. 28p. (Discussion paper no.805).

7.115 TEUBAL, Morris and STEINMUELLER, Edward
The introduction of a major new technology: externalities
and government policy. Draft. Jerusalem: The Maurice Falk
Institute for Economic Research in Israel, 1980. 34p. Mimeo.
(Discussion paper no.805).

7.116 TIDSELL, C.A.
Science and technology policy: priorities of governments.
London: Chapman and Hall, 1981.222p.

7.117   VAUGHAN, William J., RUSSELL, Clifford S. and
        COCHRANE, Harold C.
Government policies and the adoption of innovations in the
integrated iron and steel industry.   Washington   D.C.:
Resources for the Future, 1974. 458p.

7.118 WALKER, W.B.
Direct government aid for industrial innovation in the U.K.
Report to TNO/Holland. 2nd draft. Brighton: SPRU, 1976. 71
leaves. Mimeo.

7.119 WELLES, John G. and others
Contract research and development adjuncts of Federal
agencies: an exploratory study of forty organizations.
Denver: Denver Research Institute, 1969. 400p.

7.120 WHISTON, Tom
U.S. regulatory control and innovatory response in the
automobile industry: a view from Europe. Executive summary.
Brighton: SPRU, February 1981. 27 leaves. Mimeo.

7.121 WILSON, R.W., ASHTON, P.K. and EGAN, T.P.
Innovation, competition and government policy in the
semi-conductor industry. Lexington, Mass.: Lexington Books,
1980. 219p.

7.122 WORKSHOP ON GOVERNMENT POLICIES TOWARDS INNOVATION
       IN SMALL AND MEDIUM SIZED FIRMS
Organised by the Six Countries Programme on Aspects of
Government   Policies Toward Technological Innovation in
Industry. The Hague, November 21-22, 1977. Portfolio of 12
loose papers.

7.b National Policies For Technical Change & Innovation

7.123 AMBASSADE DE FRANCE
Research and Development in France. London: Ambassade de
France Service Scientifique, 1974. 36p.

7.124 AUSTRALIA - Australian Science and Technology
       Council (ASTEC)
Science and technology in Australia 1977-1978: a report to
the Prime Minister by ASTEC. 2 Vols. Canberra: Australian
Government Publishing Service, 1978-1979. Vol.1A - 145p.
Vol.1B - 131p. Vol.2 - 679p.

7.125 AUSTRALIA - Bureau of Industry Economics
Australian industrial development - some aspects of
structural change. Canberra: Australian   Government
Publishing Service, 1979. 127p. Pbk. (Research Report 2).

7.126 AUSTRALIA - Committee of Inquiry into Technological
       Change in Australia
Technological change in Australia. Report of the Committee
of inquiry into Technological Change in Australia, 1980. 4
Volumes. Canberra: Australian   Government   Publishing
Service, 1980. Pbk.

7.127 AUSTRALIA - Committee to Advise on Policies for
       Manufacturing Industries
Policies for development of manufacturing industry: a green
paper. Vol.I (report), Vol.II (statistics), Vol.IV.
Canberra: Australian   Government   Publishing   Service,

1975-1976. 3 Volumes.

7.128 AUSTRALIA - Department of Science and Technology
Bases for science and technology policy. By Geoff McAlpine and Rod Badger. Canberra: Department of Science and Technology, 1981. 45p. Mimeo.

7.129 AUSTRALIA - Department of Science and Technology
Science and technology statement, 1981-82. A statement on the Commonwealth Government sector prepared by the Department of Science and Technology on the basis of information provided by agencies of the Commonwealth Government, 23 October 1981, by the Minister for Science and Technology, David Thomson. Canberra: Australian Government Publishing Service, 1982. 145p.

7.130 CANADA - Department of Industry, Trade and Commerce,
        CAD/CAM Technology Advancement Council
Strategy for survival: issues and recommendations concerning the implementation and impact of CAD/CAM. Technology Advancement Council Secretariat, 1980. Individually paginated. Mimeo.

7.131 CANADA - Economic Council of Canada
Science, technology and innovation. By Andrew H. Wilson. Special Study No.8. Ottawa, Queen's Printer, 1968.

7.132 CANADA - Ministry of State for Science and
        Technology
R & D, innovation and economic growth: a review paper from a government perspective. By B. Bhaneja. Ottawa: Ministry of State for Science and Technology, 1981. 93p. Mimeo.

7.133 CANADA - Science Council of Canada
Forging the links: a technology policy for Canada. Ottawa, Canada: Ministry of Supply and Services, Canada, 1979. 72p. Pbk. (Science Council of Canada Report 29).

7.134 CANADA - Science Council of Canada
Address by Stuart L. Smith, Chairman, Science Council of Canada, to the AnnualGeneral Meeting of the Canadian Advanced Technology Association. Ottawa: Science Council of Canada, 1982. 13 leaves. Mimeo.

7.135 CONSERVATIVE PARTY
Proposals for a Conservative information technology policy. Provisional draft report, 1979. 57 leaves.

7.136 COOPER, Charles M.
Policy interventions for technological innovation in

developing countries. Revised and edited version. Brighton: SPRU, September 1980. 56p. Mimeo.

7.137 DALY, Anne and PRAIS, S.J.
Industrial policies and innovation: (a) The machine tool industry, (b) Types of policy. National Institute of Economic and Social Research, 1981. 26p. & 7p. Mimeo (Discussion Paper No.39 - Industry Series No. 1).

7.138 DALY, John Charles et al
Government regulation: where do we go from here? A round table held 19 December 1977 and sponsored by the American Enterprise Institute for Public Policy Research. Washington, D.C.: A.E.I., 1977. 41p.

7.139 DAVIES, D.S.
Technologists as national policy makers. The Fourteenth Maurice Lubbock Memorial Lecture. Egham, Surrey: Maurice Lubbock Memorial Fund, 1981. 18p.

7.140 DECISION SYSTEMS INC.
Canadian petrochemicals - what next: a background and planning paper on the petrochemical industry. Prepared for: Technology Assessment Division, Ministry of State for Science and Technology, Government of Canada. Toronto: Decision Systems, Inc., 1976. 319p.

7.141 DIEBOLD, Williams
Industrial policy as an international issue. New York: McGraw Hill, 1980. 305p. (1980s Project/Council on Foreign Relations).

7.142 EDWARDS, F.A.
R & D investment decision making process: a framework for policy formulation. Canada: Ministry of State for Science and Technology, 1974. 52 leaves. Mimeo.

7.143 EIRE - Ministry for Industry and Commerce
Science and Irish economic development. 2 Vols. Dublin, Stationery Office, 1966.

7.144 EIRE - National Board for Science and Technology budget
State investment in science and technology in 1980: analysis and commentary. Dublin: National Board for Science and Technology, 1980. 78p.

7.145 EIRE - National Science Council
Research and development in Ireland, 1967. By Diarmuid Murphy. Dublin: Stationery Office, 1969. 49p.

7.146   EIRE - National Science Council
Research and development in Ireland, 1969.   By   Diarmuid
Murphy and Donal O'Brolchain. Dublin: Stationery Office,
1971. 55p.

7.147   EIRE - National Science Council
Science policy formulation and resource allocation.   Dublin:
Stationery Office, 1972. 25p.

7.148   EIRE - National Science Council
Ireland: background report on science and technology.
Compiled by Diamuid Murphy. Dublin: Stationery Office,
1972. 75p.

7.149   EIRE - National Science Council
Science, technology and industry in Ireland.   By   Charles
Cooper and Noel Whelan. Dublin: Stationery Office, 1973.
43p.

7.150   EIRE - National Science Council
Studies in Irish science policy. Dublin: Stationery Office,
1973. 50p.

7.151   EIRE - National Science Council
Science Policy: some implications for Ireland.   Dublin:
Stationery Office, 1973. 20p.

7.152   EIRE - National Science Council
Research and development in Ireland, 1971. By Diamuid Murphy
and Michael Fitzgerald. Dublin: Stationery Office, 1973.
56p.

7.153   EIRE - National Science Council
Research and development in Ireland, 1975. By Conor Maguire
and Diarmuid Murphy. Dublin: Government Publications Sales
Office, 1977. 58p.

7.154   GERMANY - Federal Minister for Research and
         Technology
Sixth report of the Federal government on research.   Bonn:
Federal Minister for Research and Technology, 1980. 93p.

7.155   GERMANY - Federal Ministry for Research and
         Technology
Research and technology in the service of society.   Bonn:
Bundesministerium fur Forschung und Technologie, 1975. 21p.

7.156   GERMANY - Federal Ministry for Research and
         Technology
Research and research promotion in the Federal Republic of

Germany. Bonn: Federal Ministry for Research and Technology, 1981. 52p. Mimeo.

7.157 GERMANY - Ministry for Education and Science
Report ot the Federal Government on research. Bonn: Bonner Universitats - Buchdruckerei, 1972. 234p.

7.158 GIBBS, Roger
Industrial policy in more successful economies - Japan. London: NEDO, 1980. 67p. (Discussion paper 7).

7.159 GRANT, Paul and MACDONALD, Stuart
Innovation and the governance of technological change in Australia. Brisbane: Science Policy Research Centre, Griffith University, 1979. 66p. (Science Policy Research Centre, occasional papers no. 3).

7.160 GRAVES, Laurie E. and SCHLIE, Theodore W.
Federal incentives for innovation. Research and development incentives program publications abstracts 1972-1974. Denver: Denver Research Institute, 1975. 94p.

7.161 GREAT BRITAIN
Report ot the working party on the experimental manipulation of the genetic composition of micro-organisms. London: HMSO, 1975. 23p. Command 5880.

7.162 GREAT BRITAIN
First report of the Genetic Manipulation Advisory Group. London: HMSO, 1978. 77p. Command 7215.

7.163 GREAT BRITAIN
Biotechnology. London: HMSO, 1981. 11p. (Cmnd 8177).

7.164 GREAT BRITAIN - Advisory Board for the Research
       Councils
Energy research: the Research Council's contribution: report ot the Committee on Energy Research.   London: A.B.R.C. Dept. of Education and Science, 1974. 34p.

7.165 GREAT BRITAIN - Advisory Council for Applied
       Research and Development
Industrial innovation. London: HMSO, 1978. 43p.

7.166 GREAT BRITAIN - Advisory Council for Applied
       Research and Development
The applications of semiconductor technology. London: HMSO, 1978. 31p.

7.167 GREAT BRITAIN - Advisory Council for Applied

Research and Development
Technological change: threats and opportunities for the
United Kingdom. London: HMSO, 1979. 38p.

7.168   GREAT BRITAIN - Advisory Council for Applied
        Research and Development
Joining and assembly: the impact of robots and automation.
London: HMSO, 1979. 43p.

7.169   GREAT BRITAIN - Advisory Council for Applied
        Research and Development
Exploiting invention. London: ACARD, 1980. 12p. Mimeo.

7.170   GREAT BRITAIN - Advisory Council for Applied
        Research and Development
R & D public purchasing. London: HMSO, 1980. 44p.

7.171   GREAT BRITAIN - Advisory Council for Applied
        Research and Development
Biotechnology: Report of a Joint Working Party.   London:
HMSO, 1980. 63p.

7.172   GREAT BRITAIN - Advisory Council for Applied
        Research and Development
Computer aided design and manufacture. London: HMSO, 1980.
36p.

7.173   GREAT BRITAIN - Cabinet Office/Advisory Council for
        Applied Research and Development
Facing international competition: the impact on product
design ot standards, regulations, certification and
approvals. London: HMSO, 1982. 108p.

7.174   GREAT BRITAIN - Central Advisory Council for Science
        and Technology
Technological innovation in Britain. Sir Solly Zuckerman,
Chairman. London, HMSO, 1968.

7.175   GREAT BRITAIN - Council for Scientific Policy and
        University Grants Committee
A report ot a joint working group on computers for research.
B.H. Flowers, Chairman. London, HMSO, 1966. Cmnd.2883.

7.176   GREAT BRITAIN - Department of Education and Science
Microelectronics in education: a development programme for
schools and colleges. London: DES, 1979. 7p.

7.177   GREAT BRITAIN - Department of Industry
The economics ot industrial subsidies. Edited   by   Alan
Whiting. Papers and proceedings of the conference on the

Economics of Industrial Subsidies held at the Civil Service College, Sunningdale, February 1976. London: HMSO, 1976. 270p.

7.178 GREAT BRITAIN - Department of Industry
Industrial efficiency and the role of government. Edited by Collette Bowe. London: HMSO, 1977. 282p.

7.179 GREAT BRITAIN - Department of Industry
Report or the National Committee on Computer Networks.
Presented to the Secretary of State for Industry by J. Howlett. London: Department of Industry, 1978. 121p.

7.180 GREAT BRITAIN - Department of Industry
Japan: technology and industry by H.A.J. Prentice. London: Department of Industry, 1979. 7p.

7.181 GREAT BRITAIN - Department of Industry National Engineering Laboratory
Automated small-batch production: technical study.   East Kilbride: National Engineering Laboratory, 1978. 2 volumes (269p., 138p.).

7.182 GREAT BRITAIN - Department of Industry, Overseas Technical Information Unit
Japan, prosperity from technology, by F.D. Marshall. London: Department of Industry, 1981. 40p.

7.183 GREAT BRITAIN - Department of Industry/Shell UK Ltd.
Helping small firms start up and grow: common services and technological support. Final report. London: HMSO, 1982. 64p and Appendices. Mimeo.

7.184 GREAT BRITAIN - Department of Trade and Industry
The impact of foreign direct investment on the United Kingdom. By M.D. Steuer et al. London: HMSO, 1973. 217p.

7.185 GREAT BRITAIN - House of Commons Education, Science and Arts Committee
Biotechnology: interim report on the protection of the research base in biotechnology. Sixth report from the House of Commons Education, Science and Arts Committee, Session 1981-82. London: HMSO, 1982. 56p. (HC-289).

7.186 GREAT BRITAIN - House of Lords Select Committee on the European Communities
Biomolecular engineering. 4460/80. Session 1979/80, 66th Report. Proposal for a multiannual Community Programme of research and development in biomolecular engineering (indirect action 1981-1985) with Minutes of Evidence.

London: HMSO, 1980. 48p.

7.187 GREAT BRITAIN - Ministry of Technology
Productivity of the national aircraft effort: report of a
committee appointed by the Minister of Technology and the
President of the Society of British Aerospace Companies,
under the Chairmanship of St. John Elstub. London: HMSO,
1969. Not paginated.

7.188 GREAT BRITAIN - National Economic Development
       Corporation
NRDC: finance for innovation. London: NRDC. No date. 5p.

7.189 GREAT BRITAIN - National Economic Development
       Council
Imported manufactures: an enquiry into competitiveness.
London: HMSO, 1965. 40p.

7.190 GREAT BRITAIN - National Economic Development
       Council
Industrial strategy. Separate reports from 19 industries.
London: National Economic Development Office, 1976.

7.191 GREAT BRITAIN - National Economic Development
       Council
Paper and Board Sector Working Party Progress Report.
London: NEDO, 1978. 30p. Mimeo.

7.192 GREAT BRITAIN - National Economic Development
       Council
Industrial strategy: Electronic Consumer Goods Sector
Working Party. Summary note by the Chairman. London: NEDO,
1979. 7p. (NEDC(79)16).

7.193 GREAT BRITAIN - National Economic Development
       Council
Industrial strategy - Electronic Components Sector Working
Party, progress report 1979. London: NEDC, 1979. 14p.

7.194 GREAT BRITAIN - National Economic Development
       Council
Industrial strategy - office machinery sector working party
progress report 1979. London: NEDC, 1979. 12p.

7.195 GREAT BRITAIN - National Economic Development
       Council
Radio communications, radar and navigational aids sector
working party progress report, 1979. London: NEDC, 1979.
11p.

7.196 GREAT BRITAIN - National Economic Development
       Council
Electronic Components Sector Working Party Progress Report
1980. London: NEDO, 1980. 11p.

7.197 GREAT BRITAIN - National Economic Development
       Council
Man-made fibre production sector working party progress
report 1980. London: NEDC, 1980. 17p. Mimeo.

7.198 GREAT BRITAIN - National Economic Development
       Council
Industrial trends and prospects. London: NEDC, 1981. 24
leaves. Mimeo. (NEDC(81)6).

7.199 GREAT BRITAIN - National Economic Development
       Council
Report to the National Economic Development Council by the
Electronic Components Sector Working Pary. London: NEDO,
1981. 18p. Mimeo. (MHL No. 364)

7.200 GREAT BRITAIN - National Economic Development
       Council
Textile machinery sector working party progress report.
London: NEDC, 1981. 6p.

7.201 GREAT BRITAIN - National Economic Development
       Council, Wool Textile EDC
Wool textile EDC, progress report. London: NEDO, 1981. 12p.

7.202 GREAT BRITAIN - National Economic Development Office
Production planning and control: a report on the Mechanical
Engineering Economic Development Committee Conference, 9th
June, 1966. London, HMSO, 1966.

7.203 GREAT BRITAIN - National Economic Development Office
Printing in a competitive world. Report of the Printing &
Publishing EDC's Joint Mission to printing firms in five
countries. London: HMSO, 1970. 59p.

7.204 GREAT BRITAIN - National Economic Development Office
The plastics industry and its prospects: a report of the
Plastics Working Party of the Chemicals EDC. London: HMSO,
1972. 147p.

7.205 GREAT BRITAIN - National Economic Development Office
Reclaiming the 70's: the future for the low-cost woollen
sector. By Atkins Planning in association with N.V.
Berenschot. London: NEDO, 1972. 89p.

7.206 GREAT BRITAIN – National Economic Development Office
<u>Industrial review to 1977: textiles.</u> London: NEDO, 1974.
140p.

7.207 GREAT BRITAIN – National Economic Development Office
<u>Low-cost work aids for the clothing and garment industries.</u>
A report prepared by Inbucon/AIC Management Consultants Ltd.
for the E.D.C. for the Clothing Industry. London: NEDO,
1974. 145p.

7.208 GREAT BRITAIN – National Economic Development Office
<u>Prospects for the plastics packaging industry: a report</u>
<u>based on a study of past performance, future investment</u>
<u>requirements and constraints to profitable growth.</u> London:
NEDO, 1976. 54p.

7.209 GREAT BRITAIN – National Economic Development Office
<u>The microelectronic industry: a report by the Electronic</u>
<u>Components Sector Working party.</u> London: NEDO, 1978. 14p.

7.210 GREAT BRITAIN – National Economic Development Office
<u>Product design: a report carried out for the National</u>
<u>Economic Development Council.</u> By K.G. Corfield. London:
NEDO, 1979. 94p.

7.211 GREAT BRITAIN – National Economic Development Office
<u>The tripartite approach to industrial recovery: a guide for</u>
<u>financial institutions.</u> London: NEDO, 1980. 22p.

7.212 GREAT BRITAIN – National Economic Development Office
<u>Biotechnology.</u> Report of the Economic Development Committee
for the Chemical Industry, Specialised Organics Sector
Working Party. London: NEDO, 1980. 7 leaves and Appendix.
Mimeo.

7.213 GREAT BRITAIN – National Economic Development Office
<u>Computer manpower in the '80s: the supply and demand for</u>
<u>computer related manpower to 1985.</u> Final   report of the
Manpower Sub-Committee of the Electronic Computers SWP, by
Alan Anderson and Anton Herslab. London: HMSO, 1980. 236p.

7.214 GREAT BRITAIN – National Economic Development Office
<u>The microelectronics industry.</u> A progress report by the
Electronic Components Sector Working Party. London: NEDO,
1980. 12 leaves. Mimeo.

7.215 GREAT BRITAIN – National Economic Development Office
<u>Industrial performance: R & D and innovation.</u> EDC/SWP work
in 1980/1. London: NEDO, 1981. 27p. Mimeo. (NEDC(81)50).

7.216 GREAT BRITAIN - National Economic Development Office
<u>Industrial performance: trade performance and marketing.</u>
EDC/SWP work in 1980/81. London: NEDO, 1981. 28p. Mimeo.
(NEDC(81)41).

7.217 GREAT BRITAIN - National Economic Development Office
<u>Industrial policies in Europe: a study of policies pursued</u>
<u>in European countries and the EEC and their implications for</u>
<u>the UK.</u> London: NEDO, 1981. Individually paginated. Mimeo.
(NEDC(81)51).

7.218 GREAT BRITAIN - National Economic Development
       Office, Chemical EDC
<u>Chemicals: contraction or growth?</u> London: HMSO, 1981. 25p.

7.219 GREAT BRITAIN - National Economic Development
       Office, Chemicals EDC, Inorganics Sector Group
<u>Industrial review: inorganic chemicals.</u> London: NEDO, 1981.
27p.

7.220 GREAT BRITAIN - National Economic Development
       Office, Electronic Capital Equipment Sector Working
       Party
<u>Real-time software R & D in the U.K.: a survey and</u>
<u>recommendations for action.</u> London: NEDO, 1982. 12 leaves.
Mimeo.

7.221 GREAT BRITAIN - National Economic Development
       Office, Electronic Computers Sector Working Party,
       Manpower Sub-Committee
<u>Final report: follow up of recommendations.</u> London: NEDO,
1981. 12p. Mimeo.

7.222 GREAT BRITAIN - National Economic Development
       Office, HVACR Equipment Sector Working Party
<u>Microprocessor application in HVACR products.</u> Prepared for
the Heating, Ventilating, Air Conditioning and Refrigeration
Equipment Sector Working Party by Anthony J. Zeiling, Policy
Studies Institute. London: NEDO, 1981. 16p.

7.223 GREAT BRITAIN - National Economic Development
       Office, Joint Textile Committee
<u>Trends in textile technology.</u> London: NEDO, 1976. 12p.

7.224 GREAT BRITAIN - National Economic Development
       Office, Process Plant EDC
<u>The British process plant industry - achievements.</u> London:
NEDO, 1981. No pagination.

7.225 GREAT BRITAIN - National Engineering Laboratory

Technology transfer: the experience of the National
Engineering Laboratory. By M.M. Daniel and A.L.M.
McClintock. Paper presented at the seminar on: Transfer
processes in technical change, University of Stirling, June
1977. Glasgow: N.E.L., 1977. 13p.

7.226 GREAT BRITAIN - Office of the Minister for Science
Report of the Committee on the Management and Control of
Research and Development. Chairman: Sir Solly Zuckerman.
London, HMSO, 1961.

7.227 GREAT BRITAIN - Science Research Council
Selectivity and concentration in support of research.
London, SRC, 1970.

7.228 GREAT BRITAIN - Science Research Council
Report on superconducting A.C. generators, by a working
party of the Electrical and Systems Engineering Committee.
London: SRC, 1972. 23p.

7.229 GREAT BRITAIN - Science Research Council
The coordinated programme of research in distributed
computing systems. Annual Report, September 1978 - September
1979. London: The Computing Science Committee, Science
Research Council, 1979. 132p. Mimeo.

7.230 GREAT BRITAIN - Secretary of State for Prices and
       Consumer Protection
A review of monopolies and mergers policy: a consultative
document. London: HMSO, 1978. 161p. (Cmnd. 7198).

7.231 GREAT BRITAIN - Select Committee on Science and
       Technology
Industry and scientific research. Session 1975-76 (Science
Sub-Committee), Wednesday, 17 March 1976. Minutes of
evidence. London: HMSO, 1976. pp.251-264.

7.232 GREAT BRITAIN - Select Committee on Science and
       Technology
Session 1976-77 (Japan Sub-Committee), Thursday, 14 July
1977. Minutes of evidence by Professor R.P.Dore. London:
HMSO, 1977. pp.71-88p. 105-iv.

7.233 GREAT BRITAIN - Select Committee on Science and
       Technology
Session 1977-78. Second report from.... Innovation
research and development in Japanese science based industry.
Vol.1 report. London: HMSO, 1978. xl.

7.234 GREAT BRITAIN - Select Committee on Science and

Technology, Session 1978-79
Recombinant DNA research. Second report - interim, together with the minutes of evidence taken before the genetic engineering sub-committee and appendices. London: HMSO, 1979. 267p.

7.235 HERSKOVIC, Shlomo
The import and export of technological know-how through licensing agreements in Israel, 1966-1974. Stage 1. Jerusalem: National Council for Research & Development, 1976. 83p. (NCRD 12-76).

7.236 JAPAN - Ministry of International Trade and Industry: Institute for Transfer of Industrial Technology
Cooperation through research and development. Tokyo: M.I.T.I., 1973. 7p.

7.237 JAPAN - Science and Technology Agency
1977 white paper on science and technology - summary. October 1978. N.D.: Japanese Foreign Press Center, 1978. 75p.

7.238 JAPAN - Science and Technology Agency
1978 white paper on science and technology. Tokyo: Foreign Press Center, 1979. 44p. Mimeo.

7.239 JAPAN COMPUTER USAGE DEVELOPMENT INSTITUTE
The plan for an information society: a national goal toward year 2000. Final report of the Computerization Committee. Tokyo: The Institute, 1972. 43 leaves. Mimeo.

7.240 NATIONAL ECONOMIC DEVELOPMENT OFFICE
Organising R & D: a review of Organisational Structure for Research and Development in the Pharmaceutical Industry. London: NEDO, 1972. 32p.

7.241 NATIONAL INSTITUTE OF ECONOMIC AND SOCIAL RESEARCH. Joint studies in public policy. Conference on Industrial Policy and Innovation, held on December 9 and 10, 1980 at the National Institute of Economic and Social Research, London. London: NIESR, 1980. Assorted papers. Mimeo.

7.242 NATIONAL SCIENCE FOUNDATION
The effects of international technology transfers on the U.S. economy: present understanding, and its implications for public policy and research priorities. Papers prepared for a colloquium to be sponsored by the National Science Foundation, October, 1973. 117p.

7.243 SHAPLEY, Willis H., TEICH, Albert H. and BRESLOW, Gail J.
New directions for R & D: Federal Budget - FY 1982 - industry, defense. A report prepared for the Executive Officer and the AAAS Committee on Science, Engineering and Public Policy. Washington D.C.: AAAS, 1981. 162p. (Research and Development AAAS Report VI).

7.244 SURREY, A.J. and WALKER W.
'Energy R&D-a U.K. perspective', Energy Policy, Vol.3, No.2, June 1975.

7.245 TECHNOLOGY IRELAND
Whole issue. October 1980. 84p.

7.246 UNITED STATES - Congress House Committee on Foreign Affairs: Subcommittee on National Security Policy and Scientific Developments
Science, technology and American diplomacy. The evolution of international technology. Prepared by the Science Policy Research and Foreign Affairs Division, Legislative Reference Service, Library of Congress. Washington: USGPO, 1970. 70p.

7.247 UNITED STATES - Congress House: Committee on Science and Astronautics
Government, science and public policy; a compilation of papers prepared for the seventh meeting, panel on science and technology. Washington, USGPO, 1966.

7.248 UNITED STATES - Congress House: Committee on Science and Astronautics
Panel on science and technology, twelth meeting: international science policy. January 26-28, 1971. Washington: USGPO, 1971. 373p. 92nd congress, first session.

7.249 UNITED STATES - Congress House: Committee on Science and Technology
Science policy implictions of D.N.A. recombinant molecule research. Hearings before the Subcommittee on Science, Research & Technology, March 29,30,31; April 27,28; May 3,4,5,25,26; September 7,8, 1977. Washington, D.C.: USGPO, 1977. 1293p. 95th Congress, first session.

7.250 UNITED STATES - Department of Commerce
Factors affecting the international transfer of technology among developed countries. Report of the Panel on International Transfer of Technology. Washington, USGPO, 1970.

7.251 UNITED STATES - Department of Commerce

Technology enhancement programs in five foreign countries.
By George C. Nichols et al. Springfield: NTIS, 1972. 328p.

7.252 UNITED STATES - Department of Commerce
The American computer industry in its international
competitive environment. Washington, D.C.: Department of
Commerce, 1976. 68p.

7.253 UNITED STATES - Department of Commerce
The impact of technological innovation on international
trade patterns. By Regina K. Kelly. Washington D.C.: U.S.
Department of Commerce, 1977. 23 leaves. Mimeo. (Staff
Economic Report ER-24).

7.254 UNITED STATES - Department of Commerce
Advisory Committee on industrial innovation. Final Report.
Washington D.C.: USGPO, 1979. 299p.

7.255 UNITED STATES - Department of Commerce, Technical
       Advisory Board
The role of new technical enterprises in the U.S. economy. A
report of the Commerce Technical Advisory Board to the
Secretary of Commerce, by Richard S. Morse. Washington,
D.C.: Department of Commerce, 1976. 13p. Mimeo.

7.256 UNITED STATES - Department of Labor Manpower
       Administration
Management decisions to automate. Washington, USGPC, 1965.

7.257 UNITED STATES - Federal Trade Commission
Staff report on the semiconductor industry: a survey of
structure, conduct and performance. By Douglas W. Webbink.
Washington, D.C.: USGPO, 1977. 195p.

7.258 UNITED STATES - General Accounting Office
Implications of electronic mail for the postal service's
work force. Washington D.C.: US, GAO, 1981. 57p. Mimeo.

7.259 UNITED STATES - National Academy of Engineering
Industrial innovation and public policy options: report of
a colloquium. Washington D.C.: National Academy Press,
1980. 39p. Mimeo.

7.260 UNITED STATES - National Academy of Sciences
Committee on Computer-Aided Manufacturing in 1979 - Annual
Report to the Air Force Systems Command, U.S. Air Force.
Washington, D.C.: NAS, 1980. 34 leaves. Mimeo.

7.261 UNITED STATES - National Aeronautics and Space
       Administration/IITRI Manufacturing Applications Team

Technology transfer: solving manufacturing problems through aerospace technology. Chicago, Illinois: IIT Research Institute, 1981. 10p and Appendices. Mimeo.

7.262 UNITED STATES - National Bureau of Standards
Experimental Technology Incentives Program. Overview of regulatory experiments and publications list. Washington D.C.: ETIP, 1980. 13p. Mimeo.

7.263 UNITED STATES - National Research Council
The impact of regulation on industrial innovation. By Henry G. Grabowski and John M. Vernon. Washington D.C.: N.A.S., 1979. 64p.

7.264 UNITED STATES - National Research Council, Committee on Computer-Aided Manufacturing
Portability and integration of CAD/CAM modules: definitions and measurement. A report to the Air Force Systems Command, U.S. Air Force. Washington, D.C.: NAS, 1980. 13 leaves.

7.265 UNITED STATES - National Research Council, Committee on Computer-Aided Manufacturing
Improving managerial evaluations of computer-aided manufacturing. A report to the Air Force Systems Command, U.S. Air Force. Washington, D.C.: National Academy Press, 1981. 41p. Mimeo.

7.266 UNITED STATES - National Science Foundation
Serving social objectives via technological innovation: possible near-term federal policy options. Washington D.C.: NSF, 1973. 97p.

7.267 UNITED STATES - National Science Foundation
Technological innovation and Federal Government policy: research and analysis of the Office of National R & D Assessment. Washington D.C.: N.S.F. 1976. 73p.

7.268 UNITED STATES - Office of the White House Press Secretary
The President's industrial innovation initiatives - fact sheet. Washington D.C.: The White House, 1979. 12 leaves and President's speech to Congress, 5 leaves. Mimeo.

7.269 UNITED STATES - President's Commission for a National Agenda for the Eighties
Science and Technology: promises and dangers in the eighties. Washington, D.C.: USGPO, 1980. 93p.

7.c Other Policies On Technical Change & Innovation (OECD Etc)

7.270 BLOSSIER, D.
<u>Analysis of the research and development potentials of the member states of the European Community.</u> Luxembourg: European Communities, Directorate-General for Research Science and Education (CREST), 1979. 160p. Mimeo. (CREST/1211/79-EN).

7.271 COMMISSION OF THE EUROPEAN COMMUNITIES
<u>Industrial policy in the Community:</u> memorandum from the Commission to the Council. Brussels: The Commission, 1970. 385p.

7.272 COMMISSION OF THE EUROPEAN COMMUNITIES
<u>Industrial policy and the European Community.</u> Luxembourg: The Commission, 1972. 17p.

7.273 COMMISSION OF THE EUROPEAN COMMUNITIES
<u>Objectives and instruments of a common policy for scientific research and technological development.</u> Brussels: Commission of the European Communites, 1972. 58p. (Bulletin of the European Communities Supplement 6/72).

7.274 COMMISSION OF THE EUROPEAN COMMUNITIES
<u>A science and technology policy for the European Community.</u> Symposium held in Milan, 24-26 May, 1976. Summary of the recommendations and main contributions. Brussels: The Commission, 1976. 345p.

7.275 COMMISSION OF THE EUROPEAN COMMUNITIES
<u>The common policy in the field of science and technology. Vol.1: decisions to be taken.</u> Brussels, 1977. 83p. COM (77) 283 final.

7.276 COMMISSION OF THE EUROPEAN COMMUNITIES
<u>European society faced with the challenge of new information technologies: a community response.</u> Brussels: CEC, 1979. 48p. Mimeo.

7.277 COMMISSION OF THE EUROPEAN COMMUNITIES
<u>FAST: Sub-programme bio-society - research activities.</u> Luxembourg: CEC, Directorate-General for Research, Science and Education, 1980. 53p. Mimeo.

7.278 COMMISSION OF THE EUROPEAN COMMUNITIES
<u>Research-development in the European community: towards a new phase in the common policy.</u> Conference held at

Strasbourg, 20-22 October 1980. Strasbourg: CEC, 1980. Assorted papers. Mimeo.

7.279 COMMISSION OF THE EUROPEAN COMMUNITIES
Science and technology policy: advisory committees for the common science and technology policy. Luxembourg: Commission of the European Communities, 1980. 302p. (EUR 6745 EN).

7.280 COMMISSION OF THE EUROPEAN COMMUNITIES
A research and development strategy to meet the needs of the 1980's. Brussels: CEC, 1981. 3p. Mimeo. (Information Note No.17)

7.281 COMMISSION OF THE EUROPEAN COMMUNITIES
Relations between technology, capital and labour. Research Seminar of the Commission of the European Communities, September 3/4, 1981. Brussels: The Commission, 1981. Assorted papers. Mimeo.

7.282 COMMISSION OF THE EUROPEAN COMMUNITIES
Scientific and technical research and the European Community - proposals for the 1980's. Brussels: The Commission, 1981. 20p.

7.283 COMMISSION OF THE EUROPEAN COMMUNITIES
Notes on three French biotechnology reports. A review paper by Ken Sargeant. Brussels: CEC, 1981. 47p. (FAST).

7.284 COMMISSION OF THE EUROPEAN COMMUNITIES
Eleventh report on competition policy. Brussels/Luxembourg 1982. 194p.

7.285 COMMISSION OF THE EUROPEAN COMMUNITIES - Directorate
         of General Information, Market and Innnovation
Innovations from Community Research. Luxembourg: CEC, 1980. 13p.

7.286 COMMISSION OF THE EUROPEAN COMMUNITITES
Opinions and standpoint of the institutions and of the professionl organzations of the Community concerning the Commission's memorandum to the Council on the Industrial policy of the Community. Brussels: Commission of the European Communities, 1971. Not paginated.

7.287 COMMISSION OF THE EUROPEAN COMMUNITITES
Objectives, priorities and resources for a common research and development policy. (Communication of the Commission to the Council). Brussels: CEC 1975. 28p. Mimeo (COM (75) 535 final.

7.288   CONFEDERATION OF BRITISH INDUSTRY - Europe Committee
Industrial policy in the European Community - reappraisal
and priorities. London: C.B.I., 1976. 33p.

7.289   DANZIN, A.
Science and the second renaissance of Europe.   Oxford:
Pergamon Press for Commission of the European Communities,
1979. 124p.

7.290   ECONOMIC COMMISSION FOR EUROPE
Report of the first session of Senior Advisers to ECE
government on science and technology, 11-14 December, 1972.
Geneva: U.N. Economic and Social Council, 1972, 16p. Mimeo.

7.291   ECONOMIC COMMISSION FOR EUROPE
Factors of growth and investment policies: an international
approach. Seminar on factors of growth and investment
policies: an international approach. Budapest, 1976, 112p.

7.292   EUROPEAN COMMUNITIES
Public expenditure on Research and Development in the
Community countries: analysis by objectives, 1969-1973.
First report from the Sub-Committee 'Statistics' to the
Committee on Scientific and Technical Research (Crest).
Luxembourg: Office for Official Publications of the
European Communities, 1974. Mimeo.

7.293   EUROPEAN COMMUNITIES
Public expenditure on research and development, 1974-1976.
Luxembourg: European Communities, 1976. 131p.

7.294   EUROPEAN COMMUNITIES - Statistical Office
Government financing of research and development, 1970-1977.
Luxembourg: European Communities, 1977. 117p.

7.295   EUROPEAN ECONOMIC COMMUNITY
Working document of the Study Group on Research and
Development Policy (Section for Energy and Nuclear
Questions) on the organization and management of community R
& D. Part 8: United Kingdom. Brussels, October 1978. 25p.
Mimeo.

7.296   EUROPEAN FEDERATION OF BIOTECHNOLOGY
A community strategy for European biotechnology.   Workshop,
Oberursel, 27-30 September 1981. Frankfurt: European
Federation of Biotechnology, 1982. Vol. 1 - 142 leaves. Vol.
2 - 147 leaves. Mimeo. (Fast - Biosociety Project C. 1.1).

7.297   EUROSTAT
Government financing of research and development, 1970 -

1979. Luxembourg: The Community, 1980. 187p.

7.298 KRAMISH, Arnold
Survey of projects on the effect of research and
development expenditure investments in the economy. Paris,
OECD, 1961. SG/SPG/61.4

7.299 KRAMISH, Arnold
The contribution of research in attaining the 50% target
increase in GNP recommended by the OECD council. Paris,
OECD, 1962. SG/SPG/62.4

7.300 KRAMISH, Arnold
Research and development in the Common Market vis-a-vis the
U.K., U.S., and U.S.S.R. Santa Monica: Rand Corp., 1963.
P-2742

7.301 KRONZ, H. and GREVINK, H.
Information management: technology pattern in the EEC.
Luxembourg: Commission of the European Communities, 1976.
300p. EUR 5530 e.

7.302 LAYTON, Christopher
European advanced technology: a programme for integration.
London, Allen & Unwin for PEP, 1969.

7.303 ORGANISATION FOR ECONOMIC CO-OPERATION AND
        DEVELOPMENT
Science, economic growth and government policy. By Chris-
topher Freeman, Raymond Poignant and Ingvar Svennilson.
Paris, 1963.

7.304 ORGANISATION FOR ECONOMIC CO-OPERATION AND
        DEVELOPMENT
The residual factor and economic growth. Paris, OECD, 1964.
279p. (Study Group in the Economics of Education).

7.305 ORGANISATION FOR ECONOMIC CO-OPERATION AND
        DEVELOPMENT
Contract research institutes in Norway. By O. Elsrud. Paris,
1965.

7.306 ORGANISATION FOR ECONOMIC CO-OPERATION AND
        DEVELOPMENT
Ministerial meeting on science: Government and technical
innovation. Paris, 1966.

7.307 ORGANISATION FOR ECONOMIC CO-OPERATION AND
        DEVELOPMENT
Ministerial meeting on science: the role of government in

<u>stimulating technical innovation.</u>   Interim   committee
reference papers. Paris, 1966.

7.308   ORGANISATION FOR ECONOMIC CO-OPERATION AND
        DEVELOPMENT
<u>Gaps in technology: general report.</u> Paris, 1968.

7.309   ORGANISATION FOR ECONOMIC CO-OPERATION AND
        DEVELOPMENT
<u>Gaps in technology: scientific instruments.</u> Paris, 1968.

7.310   ORGANISATION FOR ECONOMIC CO-OPERATION AND
        DEVELOPMENT
<u>Gaps in technology between member countries: man-made
fibres.</u> Paris, 1968. CMS(68)13.

7.311   ORGANISATION FOR ECONOMIC CO-OPERATION AND
        DEVELOPMENT
<u>Gaps in technology: electronic computers.</u> Paris, 1969.

7.312   ORGANISATION FOR ECONOMIC CO-OPERATION AND
        DEVELOPMENT
<u>Gaps in technology: non-ferrous metals.</u> Paris, 1969.

7.313   ORGANISATION FOR ECONOMIC CO-OPERATION AND
        DEVELOPMENT
<u>Gaps in technology: pharmaceuticals.</u> Paris, 1969.

7.314   ORGANISATION FOR ECONOMIC CO-OPERATION AND
        DEVELOPMENT
<u>Gaps in technology: plastics.</u> Paris, 1969.

7.315   ORGANISATION FOR ECONOMIC CO-OPERATION AND
        DEVELOPMENT
<u>Gaps in technology: analytical report.</u> Comparisons between
member countries in education, research & development,
technological innovation, international economic exchanges.
Paris, OECD, 1970.

7.316   ORGANISATION FOR ECONOMIC CO-OPERATION AND
        DEVELOPMENT
<u>Productivity and economic planning.</u> Paris, OECD, 1970. 325p.

7.317   ORGANISATION FOR ECONOMIC CO-OPERATION AND
        DEVELOPMENT
<u>Transfer of technology: the role of national R & D in the
transfer of technology.</u> By Daniel Shimshoni. Paris, OECD,
1970. 10p. Mimeo.

7.318   ORGANISATION FOR ECONOMIC CO-OPERATION AND

DEVELOPMENT
NC machine tools: their introduction in the engineering
industries. Paris, 1970.

7.319   ORGANISATION FOR ECONOMIC CO-OPERATION AND
        DEVELOPMENT
The conditions for success in technological innovation.
Paris, OECD, 1971. 169p.

7.320   ORGANISATION FOR ECONOMIC CO-OPERATION AND
        DEVELOPMENT
Problems and prospects of fundamental research in
multi-disciplinary fields: computer science. Paris: OECD,
1972. 51p. (Science Policy Studies).

7.321   ORGANISATION FOR ECONOMIC CO-OPERATION AND
        DEVELOPMENT
Changing priorities for government R & D. Paris, 1973, 468p.
An experimental study of trends in the objectives of
government R & D funding in 12 OECD member countries,
1961-1972.

7.322   ORGANISATION FOR ECONOMIC CO-OPERATION AND
        DEVELOPMENT
The slowdown in R & D expenditure and the scientific and
technical system. Paris: OECD, 1974. 148p.

7.323   ORGANISATION FOR ECONOMIC CO-OPERATION AND
        DEVELOPMENT
Foreign investment and technology transfer in the
Australian mining industry. By R.B. McKern. International
Meeting of Researchers on the Transfer of Technology by
Multinational Firms, Paris, 24-28 November, 1975. Paris,
OECD, 1975. 27 leaves and appendices. Mimeo. (Conference
Paper No.16).

7.324   ORGANISATION FOR ECONOMIC CO-OPERATION AND
        DEVELOPMENT
International technology transfers in the telecom-
munications industry. By Nicolas Jequier. International
Meeting of Researchers on the Transfer of Technology by
Multinational Firms, Paris, 24-28 November, 1975. Paris,
OECD, 1975. 23 leaves. Mimeo. (Conference Paper No.14).

7.325   ORGANISATION FOR ECONOMIC CO-OPERATION AND
        DEVELOPMENT
Policy for stimulating industrial innovation - Germany.
Paris, OECD, 1975. 49p. Mimeo.

7.326   ORGANISATION FOR ECONOMIC CO-OPERATION AND

DEVELOPMENT
The role of information in innovation. Paris, OECD, 1975.
6p. Mimeo.

7.327 ORGANISATION FOR ECONOMIC CO-OPERATION AND
DEVELOPMENT
Study of technology transfer through transnational
companies. By Luis Matos. International Meeting of
Researchers on the Transfer of Technology by Multinational
Firms, Paris, 24-28 November, 1975. Paris, OECD, 1975. 38
leaves. Mimeo. (Conference Paper No.17).

7.328 ORGANISATION FOR ECONOMIC CO-OPERATION AND
DEVELOPMENT
Transfer of technolgy for pharmaceutical chemicals -
synthesis report on the experience of five industrialising
countries. by A. Cilingiroglu. Paris, OECD, 1975. 122p.

7.329 ORGANISATION FOR ECONOMIC CO-OPERATION AND
DEVELOPMENT
Policy for stimulating industrial innovation - Netherlands.
Paris: OECD, 1975. 48p. Mimeo. Also held in French.

7.330 ORGANISATION FOR ECONOMIC CO-OPERATION AND
DEVELOPMENT
Science Resources Newsletter. Bi-annual bulletins on recent
R & D trends in OECD area, 1976 - . Paris, OECD.

7.331 ORGANISATION FOR ECONOMIC CO-OPERATION AND
DEVELOPMENT
Policies for innovation in the service sector:
identification and structure of relevant factors.   Paris,
OECD, 1977. 235p.

7.332 ORGANISATION FOR ECONOMIC CO-OPERATION AND
DEVELOPMENT
Transfer of technology by multinational corporations. Vol.1:
A synthesis and country case study. Vol.2: Background
papers presented at the International Meeting of Researchers
on the Transfer of Technology by Multinational Corporations,
OECD, 24-28 November, 1975, edited by Dimitri Germidis.
Paris, OECD, 1977. 309p + 258p.

7.333 ORGANISATION FOR ECONOMIC CO-OPERATION AND
DEVELOPMENT
Expert meeting on technology transfer and national science
and technology policies. Paris, 4-5 December, 1978. 24
separate papers held.

7.334 ORGANISATION FOR ECONOMIC CO-OPERATION AND

DEVELOPMENT
Government policies and factors influencing the innovative
capability of small and medium enterprises. Paper prepared
by the OECD Secretariat and Staff Group Strategic Surveys
TNO. May 1978. Paris, OECD, 1978. 34 leaves. Mimeo.

7.335   ORGANISATION FOR ECONOMIC CO-OPERATION AND
        DEVELOPMENT
Policies for the stimulation of industrial innovation.
Vol.1:   analytical report. Paris, OECD, 1978. 167p.

7.336   ORGANISATION FOR ECONOMIC CO-OPERATION AND
        DEVELOPMENT
Policies for the stimulation of industrial innovation.
Vol.2:   country reports - Australia, Austria, Denmark,
Finland, Ireland, Netherlands, Norway, Spain, Sweden. Paris,
OECD, 1978. 206p.

7.337   ORGANISATION FOR ECONOMIC CO-OPERATION AND
        DEVELOPMENT
Policies for the stimulation of industrial innovation.
Vol.2-1:   country reports - Canada, France, Germany, Italy,
Japan, United States, United Kingdom. Paris, OECD, 1978.
422p. Pbk.

7.338   ORGANISATION FOR ECONOMIC CO-OPERATION AND
        DEVELOPMENT
Technology transfer between East & West. Paris, OECD, 1978.
431p. + separate bibliography.

7.339   ORGANISATION FOR ECONOMIC CO-OPERATION AND
        DEVELOPMENT
The objectives of government R & D funding 1973-1977.
Germany, with selected international data. Paris: OECD,
1978. 39p. Mimeo.

7.340   ORGANISATION FOR ECONOMIC CO-OPERATION AND
        DEVELOPMENT
The objectives of government R & D funding 1972/1973 -
1976/1977. New Zealand, with selected international data.
Paris: OECD, 1978. 39p. Mimeo.

7.341   ORGANISATION FOR ECONOMIC CO-OPERATION AND
        DEVELOPMENT
Expert meeting on technology transfer and national science
and technology polices. Paris, 4-5 December 1978.

7.342   ORGANISATION FOR ECONOMIC CO-OPERATION AND
        DEVELOPMENT
The impact of electronics on industrial structures and

firms' strategies: a microeconomic approach with specific
reference to the information processing and telecommunications industries. By B. Lamborghini and C. Antonelli.
Special session (D) on the impact of microelectronics on
productivity and employment. Paris, OECD, 1979. 55p.

7.343 ORGANISATION FOR ECONOMIC CO-OPERATION AND
       DEVELOPMENT
Trends in industrial R & D in selected OECD member
countries, 1967-1975. Paris: OECD, 1979. 200p.

7.344 ORGANISATION FOR ECONOMIC CO-OPERATION AND
       DEVELOPMENT
Working Party on Information, Computer and Communications
Policy. Group of experts on economic analysis of information
activities and the role of electronics and
telecommunications technologies. Chapter III, Section 6.
Biases in the measurement of real output under con-
ditions of rapid technological change. By Henry Ergas.
Paris, OECD, 1980. 16p. Mimeo.

7.345 ORGANISATION FOR ECONOMIC CO-OPERATION AND
       DEVELOPMENT
Technical change and economic policy. Sector report - the
electronics industry, by Mick McLean. Paris, OECD, 1980.
64p. Mimeo. (1195).

7.346 ORGANISATION FOR ECONOMIC CO-OPERATION AND
       DEVELOPMENT
Technical change and economic policy. Sector report - the
fertilizers and pesticides industry, by Giovanni Rufo.
Paris, OECD, 1980. 78p. Mimeo. (1193).

7.347 ORGANISATION FOR ECONOMIC CO-OPERATION AND
       DEVELOPMENT
Policy implications of data network developments in the
OECD area. Paris: OECD, 1980. 206p. (Information Computer
Communications Policy No.3).

7.348 ORGANISATION FOR ECONOMIC CO-OPERATION AND
       DEVELOPMENT
Technical change and economic policy: science and
technology in the new economic and social context. Paris:
OECD, 1980. 117p.

7.349 ORGANISATION FOR ECONOMIC CO-OPERATION AND
       DEVELOPMENT
Information activities, electronics and telecommunications
technologies: impact on employment, growth and trade. Volume
1. Paris, OECD, 1981. 140p. (Information Computer

Communications Policy, 6).

7.350 ORGANISATION FOR ECONOMIC CO-OPERATION AND
DEVELOPMENT
Science and technology policy for the 1980's. Paris, OECD,
1981. 168p.

7.351 ORGANISATION FOR ECONOMIC CO-OPERATION AND
DEVELOPMENT
Biotechnology and government policies. Paris: OECD, 1981.
20p. Mimeo.

7.352 ORGANISATION FOR ECONOMIC CO-OPERATION AND
DEVELOPMENT
Microelectronics, productivity and employment. Paris: OECD,
1981. 290p. (Information Computer Communications Policy,
No.5).

7.353 ORGANISATION FOR ECONOMIC CO-OPERATION AND
DEVELOPMENT - Committee for Science Policy
The goals of R & D in the 1970's. Prepared by Chris Freeman
et al. Paris: OECD, 1971. 75p. Mimeo.

7.354 ORGANISATION FOR ECONOMIC CO-OPERATION AND
DEVELOPMENT - Directorate for Science, Technology
and Industry
Policy for stimulating industrial innovation - Canada. (Note
by the Secretariat). Paris, OECD, 1975. 92p. Mimeo.

7.355 ORGANISATION FOR ECONOMIC CO-OPERATION AND
DEVELOPMENT - Directorate for Scientific Affairs
Science, technology and the process of technological
innovation. R & D in OECD member countries: trends and
objectives. Paris: OECD, 1970. Mimeo.

7.356 ORGANISATION FOR ECONOMIC CO-OPERATION AND
DEVELOPMENT OBSERVER (Special supplement)
OECD's economic strategy. OECD. Ministerial Council, 3-4
June, 1980. Whole issue.

7.357 THE LABOUR PARTY
Microelectronics. London: The Labour Party, 1980. 49p. (A
Labour Party Discussion Document).

7.358 UNESCO
Principles and problems of national science policies. Paris,
1967. Science Policy Studies and Documents, 5.

7.359 UNESCO
Manual for surveying national scientific and technological

potential: collection and processing of data, management of
the "R & D" system. Paris, Unesco, 1970. Science Policy
Studies and Documents, 15.

7.360 UNESCO
National science policies in Europe: present situation and
future outlook. Paris, Unesco, 1970. Science Policy Studies
and Documents, 17.

7.361 UNESCO
World plan of action for the application of science and
technology to development: Stage II. General sector on
science and technology policies and institutions, subsectors
I & II. Unesco, 1970, 60p.

7.362 UNESCO
An introduction to policy analysis in science and techno-
logy. Paris: UNESCO, 1979. 93p. Science Policy Studies and
Documents, 46.

7.363 UNESCO
National science and technology policies in Europe and
North America, 1978: present situation and future prospects.
Paris: Unesco, 1979. 458p. Science Policy Studies and
Documents, 43. In French and English.

7.364 UNESCO
Science, technology and governmental policy: a ministerial
conference for Europe and North America (Minespol II).
Paris: Unesco, 1979. 183p. Science Policy Studies and
Documents, 44.

7.365 UNESCO
Unesco's activities in science and technology in the
European and North American region. Paris: Unesco, 1979.
49p. Science Policy Studies and Documents, 45.

7.366 UNITED NATIONS – Economic Commission for Europe
Seminar on incentives to, and obstacles arising in, the
transfer of technology among EEC countries. Madrid, 25–28
September, 1972. Geneva: United Nations Economic and Social
Council, 1972. 13p. Mimeo.

7.367 UNITED NATIONS – Economic Commission for Europe
Factors of growth and investment policies: an international
approach. Seminar on factors of growth and investment
policies: an international approach, Hungary 1976. 112p.

7.368   UNITED NATIONS INDUSTRIAL DEVELOPMENT ORGANIZATION
The establishment of an international centre for genetic
engineering and biotechnology (ICGEB).   Report of a group of
experts. Vienna: UNIDO, 1981. 43p. Mimeo.

CHAPTER 8. Technical Change And Work

8.a The Effects Of Technical Change On Job Content

8.001 A.U.E.W. - T.A.S.S.
New technology: a guide for negotiators. Richmond: AUEW-
TASS, no date. 26p.

8.002 ALLEN, A. Dale
'The impact of automation on the maintenance function - some
empirical conclusions'. Journal of Industrial Engineering,
19(8), August 1968, x-xiv.

8.003 AMERICAN FEDERATION OF LABOR AND CONGRESS OF
        INDUSTRIAL ORGANISATIONS
Labor looks at automation. Washington: AFL-CIO, 1969. 36p.

8.004 ARNOLD, E. HUGGETT, C. SENKER, P. SWORDS-ISHERWOOD,
        N. and ZMROCZEK-SHANNON, C.
'Microelectronics and women's employment'. Employment
Gazette, September 1982.

8.005 ARNOLD, Erik
'Word processing and manpower'. Infotech State of the Art
Report: Office Automation, Maidenhead, Berks. Infotech
International, February 1980.

8.006 ARNOLD, Erik and SENKER, Peter
Designing the future: the implications of CAD interactive
graphics for employment and skills in the British
engineering industry. Engineering Industry Training Board
Occasional Paper No.9. Watford: Engineering Industry
Training Board, 1982.

8.007 ARNOLD, Erik, BIRKE, Lynda and FAULKNER, Wendy
'Women and microelectronics: the case of word processors'.
Women's Studies International Quarterly, Vol.4, No.3, 1981,
pp.321-340.

8.008 ASSOCIATION OF PROFESSIONAL, EXECUTIVE, CLERICAL AND
        COMPUTER STAFF
Office technology: the trade union response. First report of

the APEX word processing working party. London: APEX, 1979. 68p.

8.009 ASSOCIATION OF PROFESSIONAL, EXECUTIVE, CLERICAL AND
       COMPUTER STAFF
Automation and the office worker. Report   of   Office
Technology Working Party. London: APEX, 1980. 68p.

8.010 ASSOCIATION OF PROFESSIONAL, EXECUTIVE, CLERICAL AND
       COMPUTER STAFF
The impact of new technology in the London and Home
Counties area. Report of a survey on new technology carried
out by the London and Home Counties area of APEX. London:
APEX, 1982. 10p. Mimeo.

8.011 ASSOCIATION OF SCIENTIFIC TECHNICAL AND MANAGERIAL
       STAFFS
Technological change, employment and the need for
collective bargaining: an ASTMS discussion paper.   London:
ASTMS, 1979. 52p.

8.012 BANKING, INSURANCE AND FINANCE UNION
New technology in banking, insurance and finance.   London:
BIFU, 1982. 32p.

8.013 BARRON, Iann and CURNOW, Ray
The future with microelectronics: forecasting the effects
of information technolgy. With   other   members   of   SPRU.
London: Frances Pinter Ltd. 1979. 243p.

8.014 BELL, Martin and TAPP, Jennifer
Automation and the structure of employment in machine
shops: a report prepared for the E.I.T.B. SPRU,   1972.   2
vols. 136p and Appendices.

8.015 BIRD, Emma
Information technology in the office: the impact on women's
jobs. Manchester: EOC, 1980. 90p and Appendices. Mimeo.

8.016 BOWEN, J.A.E.
Armageddon or Utopia? A brief survey of the impacts of
microelectronics in some sectors of the service industries.
Birmingham: University of Aston, Technology Policy Unit,
1980. 100p. (Occasional Paper, Technology Policy Unit,
University of Aston in Birmingham).

8.017 BRADBURY, Frank and RUSSELL, John
Technology, change and its manpower implications: a
comparative study of the chemical and allied products
industry in the U.K., U.S. and Japan. Staines,   Middlesex:

218

Chemical and Allied Products Industry Training Board, 1980.
Individually paginated. Mimeo.

8.018 BRADY, T. LIFF, S. SENKER, P. and BRAUN, E.
Monitoring new technology and its employment and skill
implications. Draft report to the Manpower Services
Commission. Brighton: SPRU, September 1982. Mimeo.

8.019 BRADY, T., SCOTT-KEMMIS, D., and SENKER, Peter J.
The implications of technical change for skill requirements
in the folding carton industry. SPRU. Potters Bar: The
Paper & Paper Products Industry Training Board, January
1982. 103p.

8.020 BRADY, Tim, MILES, Ian with HOFFMAN, Kurt
New (information) technology and social change in the
United Kingdom: the first third of Information Technology
Year. Changement Social et Technologie en Europe Bulletin
d'Information, No.6, Mai 1982, pp.69-95.

8.021 BRAUN, E. and SENKER, P.
New technology and employment. Brighton: SPRU, January
1982. Individually paginated. Mimeo. Report prepared for the
Manpower Services Commission.

8.022 BRIGHT, James R.
'Does automation raise skill requirements?' Harvard
Business Review 36, July/August 1958, pp.85-98.

8.023 BURKE, Patti
Clerical work and technological change. The University of
New South Wales Technology and Society Program, 1981. 37p.
(TSP Paper No.1).

8.024 CAIN, J.T.
'Microprocessors and education'. Computer, Vol.10, Part 1,
1977, pp.9-10.

8.025 CANADA - Science Council of Canada
The impact of the microelectronics revolution on work and
working. Proceedings of a Workshop sponsored by the Science
Council of Canada Committee on Computers and Communication.
Ottawa: Science Council of Canada, 1980. 73p.

8.026 CARNEGIE-MELLON UNIVERSITY - Department of
       Engineering and Public Policy and School of Urban
       and Public Affairs
The impacts of robotics on the workforce and workplace.
Preliminary draft. Pittsburgh, Penn.: Carnegie-Mellon
University, 1981. 218p. Mimeo.

8.027 CHAMOT, Dennis and BAGGETT, Joan M. (eds)
Silicon, satellites and robots: the impacts of technological change on the workplace. Washington DC: Department
for Professional Employees, 1979. 52p.

8.028 COUNCIL FOR EDUCATIONAL TECHNOLOGY FOR THE UNITED
      KINGDOM
Microelectronics: their implications for education and
training. A statement by the C.E.T. London: C.E.T, 1978.
16p.

8.029 COUNCIL FOR SCIENCE AND SOCIETY
New technology: society, employment and skill. Report of a
working party. London: CSS, 1981. 103p.

8.030 CROSSMAN, Edward R.F.W. and others
Evaluation of changes in skill-profile and job-content due
to technological change: methodology and pilot results from
the banking, steel and aerospace industries.   Report
submitted to Office of Manpower Policy, Evaluation and
Research. Berkeley, Department of Industrial Engineering and
Operations Research, University of California, 1966.

8.031 CROSSMAN, Edward R.F.W. and others
The impact of numerical control on industrial relations at
plant level: U.S.A. A report prepared for the Automation
Unit, International Labour Office. Berkeley, University of
California, Human Factors in Technology Research Group,
Department of Industrial Engineering and Operations
Research, 1968.

8.032 DAVIS, Louis E.
Discussion of the impact of automation on occupational
distribution, job content and working conditions. O.E.C.D.,
North American Joint Conference on the Manpower Implications
of Automation, Washington D.C.: December 8-10, 1964.

8.033 DYMMEL, Michael D.
'Technology in telecommunications: its effect on labour and
skills'. Monthly Labour Review, Vol.102, No.1, 1979,
pp.13-19.

8.034 GREAT BRITAIN - Department of Employment: Work
      Research Unit
A strategy for change. By Gordon Ross. Paper presented at
the ISL Conference "Word Processing - the Human Dimension",
October 30/31st 1978. London: WRU, 1978. No pagination.
Mimeo.

8.035 GREEN, Marjorie

'Automation of foreign exchange and other trading operations'. _The Banker_, Vol.131, No.622, April 1981, pp.119-123.

8.036 GRIMES, Jack D.
'How changing technology should effect computer science and engineering education'. _Proceedings COMSAC '78_, Chicago, 13-16 November 1978. pp.415-418.

8.037 HALEVI, Gideon
The role of computers in manufacturing processes. New York, Chichester: John Wiley and Sons, 1980. 502p.

8.038 HIGGINS, Peter G.
Impact of industrial robots in Australia. Second submission to Committee of Inquiry into Technological Change in Australia. Hawthorne: Mechanical Engineering Department, Swinburne College of Technology, 1979. 25 leaves. Mimeo.

8.039 HOFFMAN, Kurt
Job creation and microelectronics-based technical change: the role of software. Brighton: SPRU, August 1982. 39 leaves + tables. Mimeo.

8.040 HOPKINS, Michael and VAN DER HOEVEN, Rolph
'A model of new technology and jobs'. _Futures_, Vol. 13, No. 6, December 1981, p. 483-488.

8.041 HUGGETT, C.M. and BELL, R.M.
Microelectronics, welding and skills: a review of some trends in the engineering industry. Brighton: SPRU, September 1979. 41 leaves and tables. Mimeo.

8.042 HUGHES, James J., PERLMAN, Richard and DAS, Satya P.
'Technological progress and the skill differential'. _The Economic Journal_, Vol.91, December 1981, pp.998-1005.

8.043 HUWS, Ursula
Your job in the eighties: a woman's guide to new technology. London: Pluto Press, 1982. 127p.

8.044 INFORMATION MANAGEMENT
The impact of new technologies on publishing. Munich, London: K.G. Saur, 1980.

8.045 INGERSOLL ENGINEERS
Industrial robots: the report by Ingersoll Engineers. Vol.1: An industrial application. Vol.2: A world wide literature survey - applications R & D bibliography. East Kilbride, Glasgow: National Engineering Laboratory, 1980. Vol.1 -

separately paginated. Vol.2 - 160p. This report was contributed to by the N.E.L. and commissioned by the Department of Industry.

8.046 INSTITUTE FOR INDUSTRIAL RESEARCH AND STANDARDS
Microelectronics - implications for the Irish apparel industry. Report of an investigation conducted on behalf of the National Board for Science and Technology by the Textiles Division of the Institute for Industrial Research and Standards. Dublin: IIRS, 1980. 64p. Mimeo.

8.047 INTERNATIONAL FEDERATION OF COMMERCIAL, CLERICAL AND
       TECHNICAL EMPLOYEES
Computers and work: F.I.E.T. action programme.   Geneva: FIET, 1979. 20p.

8.048 KARAS, G. Christopher
Advanced technical education and training in plastics: an edited version of a study of the skills used in the Plastics processing industry. Carried out for the National Economic Development Office during autumn 1973. No imprint. 35p. Mimeo.

8.049 KAUFMAN, Morris, CHALLIS, A.A.L. and KARAS, G.
       Christopher
Implications of the SPRU study on the use of skills within the injection moulding sector of the plastics industry. London: RPPITB, Polymer Engineering Directorate, 1980. 29p. Mimeo.

8.050 LEEDS, Simon
'Computerising the clothing industry'.   British Clothing Manufacturer, February 1981, p.7 + p.21.

8.051 LIPSTREU, Otis and REED, Kenneth A.
Transition to automation: a study of people, production and change. Boulder, University of Colorado Press, 1964. University of Colorado Studies, Series in Business, 1.

8.052 LLOYD, P. and MILLS, S.
'Automation and industrial workers - an international study'. Personnel Review, Vol.4, Part 2, 1975, pp.41-44.

8.053 MACDONALD, Stuart and MANDEVILLE, Thomas
'Word   processors   and   employment'.   The Journal of Industrial Relations, June 1980, pp.137-148.

8.054 MILLS, Bryan E.
'Micro-processors - their impact on work and participation'. Industrial Participation, No.566, 1979, pp.15-19.

8.055 NALGO
The future with new technology: a NALGO view.   London:
NALGO, 1981. 57p.

8.056 P.A. MANAGEMENT CONSULTANTS LIMITED
The manpower implications of computerisation within the
clothing industry in Northern Ireland. Belfast: Department
of Manpower Services, 1979. No pagination. Mimeo.

8.057 PALMER, Lesley S.
Technical change and employment in banking. Presented   in
partial fulfilment of requirements for the degree of Master
of Science in History and Social Studies of Science.
Brighton: University   of   Sussex, 1980. 49 leaves +
appendices. Mimeo.

8.058 PEITCHINIS, Stephen G.
The effect of technological changes on education and skill
requirements of industry. Ottawa: Department of Industry,
Trade and Commerce, 1978. 272p.

8.059 RIJNSDORP, J.E. (ed)
I.F.A.C. workshop on case studies in automation related to
humanization of work. Oxford: Pergamon Press, 1977. 106p.
(International Federation of Automatic Control).

8.060 ROSENBROCK, H.H.
'Engineers and the work that people do'. Invited paper for
publication in: IEEE Control Systems Magazine and in:
IEEE Technology and Society. (No imprint, no date). 13p.
Mimeo.

8.061 ROSENBROCK, H.H.
'The future of control'. Automatica, Vol.13, 1977,
pp.389-392.

8.062 ROTHWELL, Roy
Technological change in textile machinery: manpower
implications in the user and producer industries. Prepared
for: NSF/BMFT Seminar "Public/Private Co-operation for
technological innovation", June 6-10, 1977, Geneva. 22
leaves. Mimeo.

8.063 SANDBERG, Ake (ed)
Computers dividing man and work: recent Scandinavian
research on planning and computers from a trade union
perspective. Stockholm: Arbetslivscentrum, 1979. 146p.

8.064 SCIBBERAS, Ed., SWORDS-ISHERWOOD, Nuala and SENKER,
       Peter

Competition, technical change and manpower in electronic capital equipment: a study of the UK minicomputer industry. Brighton: SPRU, 1978. 73p. (SPRU Occasional Paper Series No.3).

8.065 SCOTT-KEMMIS, D., BRADY, T. and SENKER, P.
The implications of technical change for skill requirements in the fibre board packing case sector. Paper and Paper Products Industry Training Board. January 1982.

8.066 SENKER, P.
Competition, technical change and skilled manpower in small engineering firms:a study of the U.K. precision press tool manufacturers. A report prepared for the Engineering Industry Training Board. Brighton: SPRU, February 1979. 88p.

8.067 SENKER, P. HUGGETT. C, BELL. M and SCIBERRAS, E.
Technological change and manpower in the U.K. toolmaking industry. Engineering Industry Training Board Research Paper No. 2. E.I.T.B., 1976. 72p.

8.068 SENKER, Peter J.
'Computerised warehouses need more than mechanical salesmen'. Mechanical Handling Survey. The Engineer, Vol.247, No.6392, 28 September 1978, pp.S17-S19.

8.069 SENKER, Peter J., SWORDS-ISHERWOOD, Nuala, BRADY, Tim and HUGGETT, Charlotte
Maintenance skills in the engineering industry: the influence of technological change. Watford: EITB, 1981. 58p. (EITB occasional paper no.8).

8.070 SHAW, B. (ed)
Computers and the educated individual. Proceedings of the Joint IBM/University of Newcastle upon Tyne Conference, 9-12 September 1975. Newcastle: University of Newcastle Computing Laboratory, 1976. 186p. Mimeo.

8.071 SOUTHERN SCIENCE AND TECHNOLOGY FORUM
Technology - employment - education. International Conference, 1979, held at the University of Southampton, September 1979. Southampton: University of Southampton, 1979. 123p.

8.072 SPRU WOMEN & TECHNOLOGY STUDIES
Microelectronics and women's employment in Britain. Based on a report for the Department of Employment and the Manpower Services Commission. Brighton: SPRU, 1982. 122p. (SPRU occasional paper series no.17).

8.073 SULTAN, Paul and PRASOW, Paul
The skill impact of automation. Los Angeles, Institute of
Industrial Relations, 1964. Reprint 136.

8.074 SWORDS-ISHERWOOD, N. and SENKER, P. (eds)
Microelectronics and the engineering industry: the need for
skills. London: Francis Pinter, 1980. 186p.

8.075 SWORDS-ISHERWOOD, Nuala and SENKER, Peter
'Automation in the engineering industry'. Labour Research,
Vol.67, No.11, November 1978, pp.230-1.

8.076 THOMAS, Graham and ZMROCZEK-SHANNON, Christine
Technology and household labour: are the times a-changing?
Paper presented at the British Sociological Association
Conference on "Gender & Society", Manchester, 5-8 April
1982. Brighton: SPRU, March 1982. 26p. Mimeo.

8.077 UNION OF COMMUNICATION WORKERS - Executive Council
Posts, telecommunications and the new technology. A commentary on the Report of the Science Policy Research Unit by
the Executive Council of the Union of Communication Workers.
London: UCW, 1980. 29p.

8.078 WALSH, V.M., TOWNSEND, J.F., SENKER, P.J. and
          HUGGETT, C.M.
Technical change and skilled manpower needs in the plastics
processing industry. Brighton: SPRU, September 1980. 105p.
(SPRU occasional paper no.11).

8.079 WALSH, Vivien, MOULTON-ABBOTT, John and SENKER,
          Peter
New technology, the Post Office and the Union of Post
Office Workers. A report prepared for the Union of Post
Office Workers. Brighton: SPRU, 1980. 74p.

8.080 ZMROCZEK, Christine and HENWOOD, Felicity
New information technology and women's employment. Brussels:
Commission of the European Communities, FAST Series, 1983.

8.b Technical Change, Employment & Unemployment

8.081 ASSOCIATION OF PROFESSIONAL, EXECUTIVE, CLERICAL AND
          COMPUTER STAFF
New technology and redundancies. London: APEX, 1980. 5p.
Mimeo.

8.082   ASSOCIATION OF SCIENTIFIC TECHNICAL AND MANAGERIAL
STAFFS
Technological change, employment and the need for
collective bargaining. An ASTMS discussion paper. London:
ASTMS, 1979.

8.083   AUTOMATION AND UNEMPLOYMENT
Papers presented at ANZAAS Symposium, Sydney 28 July 1979.
Sydney: The Law Book Co. Ltd. 1979. 164p.

8.084   BOURNER, Tom
'The impact of microelectronics on unemployment – an
economic appraisal'. Economics, Quarterly Journal of the
Economics Association, Vol. XVIII, Spring 1982, p.19–25.

8.085   BRADY, T. and LIFF, S.
Monitoring new technology and employment. Draft report to
the Manpower Services Commission. Brighton: SPRU, November
1982. Mimeo.

8.086   BRAUN, E. and SENKER, P.
New technology and employment. Brighton: SPRU, January
1982. Individually paginated. Mimeo. Report prepared for the
Manpower Services Commission.

8.087   CLUTTERBUCK, David and HILL, Roy
The re-making of work: changing work patterns and how to
capitalize on them. London: Grant McIntyre, 1981. 216p.

8.088   CONFEDERATION OF BRITISH INDUSTRY
Jobs – facing the future: a CBI staff discussion document.
London: CBI, 1980. 51p.

8.089   CURNOW, Ray
The economic impact of microprocessors on industry and
employment. Paper presented at the National Conference
Planning for Automation, 17 January 1979, Polytechnic of the
South Bank. London: The Polytechnic, 1979. pp.40–43.

8.090   EIRE, National Board for Science and Technology
Microelectronics: the implications for Ireland – an
assessment for the eighties. 2 vols. Dublin: NBST, 1981.
Vol.1 – 93p. Vol.2 – 100p. (Bibliography).

8.091   EUROPEAN TRADE UNION INSTITUTE
The impact of microelectronics on employment in Western
Europe in the 1980's. Brussels: E.T.U.I. 1979. 164p.

8.092   FOSTER, David
Innovation and employment. Oxford: Pergamon Press, 1980.

193p.

8.093 FREEMAN, C.
Science, technology and unemployment. Paper presented as a
public lecture at Imperial College of Science & Technology,
London, 9.2.82. Brighton: SPRU, March 1982. 29 leaves.
Mimeo. (To be published in revised form as no. 1 in the
series Papers in Science, Technology and Public Policy, from
Imperial College & SPRU).

8.094 FREEMAN, C. and CURNOW, R.
Technical change and employment - a review of post-war
research. Paper prepared for the Manpower Services
Commission, June 1978. Brighton: SPRU, 1978. 21p.

8.095 GERSHUNY, J.I.
Technical innovation and women's work in the EEC: a medium
term perspective. A briefing for an EEC Seminar at the Equal
Opportunities Commission, 30 May 1980. Brighton: SPRU,
April 1980. 1st draft (No pagination). Mimeo.

8.096 GERSHUNY, J.I. and MILES, Ian
Service employment: trends and prospects.   Brussels:
Commission of the European Communities, FAST Series No. 4.
(EUR7954EN.)

8.097 GERSHUNY, J.I. and MILES, Ian
The new service economy: the transformation of employment
in industrial societies. London: Francis Pinter, 1983.

8.098 GILL, Ken
'Micro-electronics and employment - a trade union view'.
Management Services, Vol.22, No.12, 1978, pp.20-21.

8.099 GREAT BRITAIN - Department of Employment
The manpower implications of micro-electronic technology. By
Jonathan Sleigh et al. London: HMSO, 1979. 110p.

8.100 GREAT BRITAIN - Manpower Services Commission
New technology and employment. By E. Braun and P. Senker.
London: Manpower Services Commission, 1982. Not paginated.

8.101 HAYWOOD, Bill
'Technological change and employment'. Clerkenwell Bulletin,
No. 2, July 1981, p. 1-2.

8.102 HESSELMAN, Linda and SPELLMAN, Ruth
Responses to the employment consequences of technological
change. London: NEDO, 1980. 23 leaves. Mimeo.

8.103 HINES, Colin and SEARLE, Graham
Automatic unemployment. London: Earth Resources Research
Ltd., 1979. 77p. Pbk.

8.104 ISCOL Ltd.
Information technology and job creation potential, phase
22. Specific study: software and the software industry. A
study carried out for the Commission of the European
Communities. Lancaster: ISCOL Ltd., 1981. 129p. Mimeo. (EEC
FAST − 82).

8.105 JAHODA, Marie and RUSH, Howard
Work, employment and unemployment: an overview of ideas and
research results in the social science literature. Brighton:
SPRU, September 1980. 70p. Mimeo. (SPRU occasional paper
no.12).

8.106 JENKINS, Clive and SHERMAN, Barrie
Computers and the unions. London: Longman, 1977. 135p.

8.107 JENKINS, Clive and SHERMAN, Barrie
The collapse ot work. London: Eyre Methuen, 1979. 181p.
Pbk.

8.108 JENKINS, Clive and SHERMAN, Barrie
The leisure snock. London: Eyre Methuen, 1981. 185p. Pbk.

8.109 KETTERINGHAM, P.J.A.
The human factor in automation. London: British Steel
Corporation, 1975. 8p.

8.110 LAYARD, Richard
Unemployment in Britain: causes and cures. London: London
School of Economics, Centre for Labour Economics, 1981. 26p.
Mimeo. (Discussion Paper No.87).

8.111 MACKINTOSH, Ian M. and JACOBS, Tom
The impact of new technology on employment. Paper presented
to the Manchester Statistical Society on 15 January 1970.
Manchester M.S.S., no date. 23p.

8.112 MALLIER, A.T. and ROSSER, M.J.
'Female employment and unemployment in a local labour
market'. Journal of Industrial Affairs, Vol.8, No.1, Autumn
1980, pp.13−22.

8.113 METRA CONSULTING GROUP LTD.
The impact of chip technology on employment and the labour
market by P.M.H. Kendall et al. Report commissioned by the
Ministry of Social Affairs, The Netherlands. London: Metra,

1979. 416p.

8.114 NEISSER, Hans P.
'Permanent technological unemployment: demand for commodities is not demand for labor', American Economic Review, Vol. 32, March 1942, p. 50-71.

8.115 NORMAN, Colin
Microelectronics at work: productivity and jobs in the world economy. Washington D.C.: Worldwatch Institute, 1980. 63p. (Worldwatch Paper 39).

8.116 PEITCHINIS, Stephen G.
Technological changes and the sectoral distribution of employment. Paper presented at the annual meeting of the Western Economic Association, 21 June, 1979. University of Calgary, Department of Economics, 1979. 27 leaves. Mimeo.

8.117 PELTU, Malcolm
'Microelectronics and unemployment'. New Scientist Vol.93, No.1290, 28 January 1982, pp.226-8.

8.118 PILORUSSO, F.
The labour displacement implication of microelectronics technology in automotive assembly plants: a case study. Submitted to: Ontario Ministry of Labour for the Microelectronics Task Force. Toronto, Ontario. (No publisher), 1981. 32 leaves. Mimeo.

8.119 ROBINS, Kevin and WEBSTER, Frank
New technology: the trade union response in the U.K. (No imprint). 37p. Mimeo.

8.120 ROSENBROCK, H.H.
Automation - economics - employment. Paper to be presented at Finnish Engineering Days Seminar, 7-8 November 1979. (No imprint). 1979. 15p. Mimeo.

8.121 ROTHWELL, Roy
Technology, structural change and manufacturing employment. Laxenburg, Austria: IIASA, December 1980. 30p. (CP-80-37). and Omega, Vol.9, No.3, 1981.

8.122 ROTHWELL, Roy and ZEGVELD, Walter
Technical Change and Employment. London: Frances Pinter, 1979. 178p.

8.123 RUSH, Howard
Automation and employment. Paper presented to the Agricultural Club, University of Reading, 14th Annual

Conference, February 6th 1980. Brighton: SPRU, January 1980. 17 leaves.

8.124 SEMA PROSPECTIVE/ISCOL
The potential of information technologies for job creation. Report on the first phase: the analysis of present data and construction of scenarios. Brussels: Commission of the European Communities, 1981. 284p. Mimeo. (FAST - 82).

8.125 SENKER, Peter J.
'Jobs murder: micros only accessories after the fact'. Electronics Times, No.54, 1 November 1979, p.18.

8.126 SENKER, Peter J.
'Barriers to automation', Development of integrated manufacturing systems, London: Institution of Production Engineers, 1980.

8.127 SHANKS, Michael
'What future for work?'. Industrial Society, Vol.62, July-August, 1979. pp.3-4.

8.128 SHOWLER, Brian and SINFIELD, Adrian
The workless state: studies in unemployment. Oxford: Martin Robertson, 1981. 267p.

8.129 SIM, R.M.
Computer aided manufacture in batch production. A background paper for the AUEW (TASS) Conference on Computer Technology and Employment, 16 September 1978. Glasgow: National Engineering Laboratory, 1978. 7p. Mimeo. (Document G).

8.130 SMITH, Joy Selby
Implications of developments in microelectronic technology on women in the paid workforce. Revised version of a paper prepared for the OECD Working Party on Information, Computer and Communications Policy. Canberra: Department of Economics, Australian National University. No date. 18 leaves. Mimeo.

8.131 SOHYO NEWS
'Development of the electronic industry and its effects upon employment'. Sohyo News, No.368, November 15, 1981. Whole issue. 16p.

8.132 STIEBER, Jack (ed)
Employment problems of automation and advanced technology: an international perspective. Proceedings of a conference held at Geneva by the International Institute for Labour Studies, 19-24 July 1964. London, Macmillan, 1966.

8.133 STONIER, Tom
'The impact of microprocessors on employment'. Employee Relations, Vol.1, Part 4, 1979, pp.27-28.

8.134 STONIER, Tom and THORNTON, Peter
'Microprocessor technology - tackling the labour problems it creates'. Iron and Steel Training, Vol.5, August 1979, pp.1-2.

8.135 STROUD, Dick and TODD, Paul
'Technological development: the long-term impact on banking'. The Banker, Vol.131, No.622, April 1981, pp.141-143.

8.136 TECHNOLOGY AND EMPLOYMENT - the impact of
       micro-electronics.
Papers from a workshop held at the Institute of Production, Aalborg University Center, 19 February 1981. Aalborg: Aalborg University Center, Institute of Production, 1981. 198p. (Industrial Development Research Series No.20).

8.137 TRADES UNION CONGRESS
Employment and technology (final report). London: T.U.C., 1979, 71p.

8.138 TRADES UNION CONGRESS - West Midlands Regional
       Council
Trade Unions and New Technology. Report of a Trade Union Conference on the Use and Impact of Microelectronics. Organized by West Midlands Regional T.U.C. and the Technology Policy Unit, Aston University, Birmingham. TPI, Aston University, 1979. 33p. Mimeo.

8.139 TRANSPORT AND GENERAL WORKERS UNION
Microelectronics: new technology, old problems, new opportunities. London: TGWU, 1979. 20p. (A TWGU Research and Education Booklet).

8.140 VOGLER, Jon
Work from waste: recycling wastes to create employment.
London: Intermediate Technology Publication/Oxfam, 1981. 396p. Pbk.

8.141 WEINSTEIN, Caroline
'Robots in industry - too few or too many?' Electronics and Power, May 1978, pp.379-383.

8.142 WEST MIDLAND T.U.C.
Trade unions and new technology. Report of a Trade Union Conference on the use and impact of microelectronics

organized by West Midlands Regional TUC and the Technology Policy Unit, Aston University. Birmingham: Aston University, 1979. 32p. Mimeo.

8.143 WILLIAMS, Bruce
Technological employment and unemployment and the implications for education. Hobart: Department of Political Science, University of Tasmania, 1979. 19p. (Public Policy Paper 14).

8.144 ZEMEN, Z.P.
The impact of computer/communications on employment in Canada: an overview of current OECD debates. Montreal: Institute for Research on Public Policy, 1979. 296p.

CHAPTER 9. <u>Bibliographies On Technical Change And Innovation</u>

9.001 AMADEO, Eduardo and ACERO, Liliana
<u>A bibliographical survey on industrial research institutes:</u>
<u>suggestions for further research.</u> Buenos Aires, 1977. 85
leaves. Mimeo.

9.002 ANDREWS, Kay, POOLE, John and WOLFF, Susan (under
      direction of Alan Mencher)
<u>Bibliography for the study of UK science and technology</u>
<u>policy with special reference to economic and industrial</u>
<u>development, part I: 1959-1973.</u> 92p. Mimeo.

9.003 BESSANT, J.R. et al
<u>The impact of microelectronics: a review of the literature.</u>
The Technology Policy Unit, University of Aston. London:
Francis Pinter 1981. 174p.

9.004 BIBLIOGRAPHY ON PROJECT SELECTION
<u>Program ot Research on the Management of Research &</u>
<u>Development - Project 10.</u> pp.159-164. Mimeo.

9.005 BRITISH COUNCIL OF PRODUCTIVITY ASSOCIATIONS
<u>Bibliography - microprocessors.</u> London: British Council of
Productivity Associations. No date. 2 leaves.

9.006 BRONSON, Leisa
<u>Cartels and international patent agreements: a selected and</u>
<u>annotated bibliography.</u> Washington, Library of Congress
Legislative Reference Service, 1943. Mimeograph.

9.007 COMMISSION DES COMMUNAUTES EUROPEENNES
<u>Social change and technology in Europe - bibliography.</u>
Bruxelles: Commission of the European Communities, 1982.
105p. (Information bulletin no. 3).

9.008 COUNCIL OF PLANNING LIBRARIANS
<u>Multinational corporations, technology transfer and the</u>
<u>developing countries: an introductory bibliography.</u> By R.D.
Steele. Council of Planning Librarians, 1975. 39p. (Exchange
bibliography 745).

9.009 DEDIJER, S. and TOMIN, U.
<u>Regional science and technology policy: an attempt at an</u>

international bibliography. Lund: Lunds Universitet, 1972. 55 leaves.

9.010 DICKSON, Keith and MARSH, John
The microelectronics revolution: a brief assessment of the industrial impact with a selected bibliography. Birmingham: Technology Policy Unit, University of Aston in Birmingham, 1978. 87p.

9.011 FUSFIELD, Herbert I.
Perspectives on U.S. industrial innovation.   Annotated bibliography by Theodore W. Schlie. New York: Center for Science and Technology Policy, New York University, 1981. 65 leaves and bibliography. Mimeo.

9.012 GOSLIN, Lewis N.
A selected annotated bibliography of R & D management. Bloomington, Indiana, Bureau of Business Research, Graduate School of Business, Indiana University, 1966.

9.013 GRAZING, Eugene E
Cost estimating for research, development, and information processing programs: an annotated bibliography. Sunnyvale, Calif., Lockheed Missiles & Space Co., 1963.

9.014 GREAT BRITAIN – Department of Industry
A strategy for industry: an annotated bibliography.   By Sylvia Waterson. London: Department of Industry, 1979. 32p. Mimeo.

9.015 GREAT BRITAIN – Department of Trade and Industry
Small firms: register of research and centres of specialised assistance. London: D.T.I. 1973. 163p.

9.016 GREAT BRITAIN – Manpower Services Commission
Research into new technology. London: MSC, 1981. No pagination. Mimeo.

9.017 GREAT BRITAIN – Work Research Unit
New technology: computers and the quality of working life – a bibliography. April 1979. London: W.R.U., 1979. 5p. (WRU Bibliography No.26).

9.018 GREAT BRITAIN – Work Research Unit
New Technology: effects of, general – a bibliography. April 1979. London: W.R.U., 1979. 5p. (WRU Bibliography No.18).

9.019 GREAT BRITAIN – Work Research Unit
New Technology: occupational effects – a bibliography. April 1979. London: W.R.U., 1979. 5p. (WRU Bibliography No.19).

9.020 GREAT BRITAIN - Work Research Unit
New Technology: employee attitudes - a bibliography. April
1979. London: W.R.U., 1979. 4p. (WRU Bibliography No.22).

9.021 GREAT BRITAIN - Work Research Unit
New Technology: job content effects - a bibliography. April
1979. London: W.R.U., 1979. 2p. (WRU Bibliography No.20).

9.022 GREAT BRITAIN - Work Research Unit
New Technology: robots - a bibliography. April 1979. London:
W.R.U., 1979. 3p. (WRU Bibliography No.27).

9.023 GREAT BRITAIN - Work Research Unit
New Technology: strategies for introduction - a
bibliography. April 1979. London: W.R.U., 1979. 2p. (WRU
Bibliography No.23).

9.024 GREAT BRITAIN - Work Research Unit
New Technology: word processing - a bibliography.   April
1980. London: W.R.U., 1979. 2p. (WRU Bibliography No.25).

9.025 GREAT BRITAIN - Work Research Unit
New Technology: work organisation effects - a bibliography.
April 1979. London: W.R.U., 1979. 3p. (WRU Bibliography
No.21).

9.026 GREAT BRITAIN - Work Research Unit
New Technology: electronic office - a bibliography.   April
1980. London: W.R.U., 1980. 5p. (WRU Bibliography No.24).

9.027 HAUSTEIN, Heinz-Dieter
Innovation glossary. Laxenburg: I.I.A.S.A., 1982. 196p.
(Working paper - 82-2).

9.028 HELLER, Terry Sovel, GILMORE, John S. and BROWNE,
       Theodore D.
Technology transfer - a selected bibliography.   Revised
edition. Washington, D.C.: National Aeronautics and Space
Administration, 1971. 175p. (NASA CR - 1724).

9.029 INSTITUTE FOR INTERNATIONAL CO-OPERATION
Scientific and technological innovation: self reliance and
co-operation: selective bibliography. Ottawa: I.I.C., 1976.
145p.

9.030 INSTITUTE OF MANPOWER STUDIES
Aspects of management development: a selective, annotated
bibliography. Sussex: IMS, 1972. 48p.

9.031 LIOBERGER, Herbert Frederick
Adoption of new ideas and practices: a summary of the

research dealing with the acceptance of technological change in agriculture, with implications for action in facilitating such change. Ames, Iowa State U.P., 1960.

9.032 METCALFE, J.S. and STUBBS, P.C.
The economics of technological change and innovation: a selected bibliography. Manchester: The University, 1973. 39 leaves. Mimeo.

9.033 NELSON, Richard R.
The economics of invention: a survey of the literature. Santa Monica: Rand Corporation, 1980.

9.034 ORGANISATION FOR ECONOMIC CO-OPERATION AND
         DEVELOPMENT
Bibliography 3: research workers. Paris, 1963.

9.035 ORGANISATION FOR ECONOMIC CO-OPERATION AND
         DEVELOPMENT
Bibliography 4: applied research. Paris, 1963.

9.036 ORGANISATION FOR ECONOMIC CO-OPERATION AND
         DEVELOPMENT
Bibliography 5: research and development. Paris, 1963.

9.037 ORGANISATION FOR ECONOMIC CO-OPERATION AND
         DEVELOPMENT
Bibliography 6: research policy programmes and selection of projects. Paris, 1963.

9.038 ORGANISATION FOR ECONOMIC CO-OPERATION AND
         DEVELOPMENT
Bibliography 12: dissemination of research results.   Paris, 1965.

9.039 ORGANISATION FOR ECONOMIC CO-OPERATION AND
         DEVELOPMENT
The integration of research policy into overall company policy: general bibliography.   European/North   American conference   on   research   management,   Monte-Carlo,   21-24 February, 1965. Paris, 1965.

9.040 PHILIPS, Glen
Technology, employment, education: a selected bibliography. Southern   Science   and   Technology   Forum,   University   of Southampton, 10 September 1979. (No imprint). 7p. Mimeo.

9.041 RETTIG, Richard A., SORG, James D. and MILWARD, H.
         Brinton
Criteria for the allocation of resources to research and development: a review of the literature.   Columbus,   Ohio:

Ohio State University Research Foundation, 1974. 143p. (RF project 3702 - final report).

9.042 ROGERS, Everett M. et al
Bibliography of the diffusion of innovations. Monticello, Illinois: Council of Planning Librarians, 1977. 234p. (Exchange Bibliography 1420, 1421 and 1422).

9.043 ROLLEFSON, Brenda
Bibliography on methodologies for evaluating research and development projects. Ottawa: Department of National Defence, 1971. 10p. (Report No. Plans 71 - 8).

9.044 ROTHMAN, Harry et al
Biotechnology: a synoptic review with selected annotated bibliography and directory of companies. Birmingham: Technology Policy Unit, University of Aston in Birmingham, 1980. 146p. Mimeo. (TPU Occasional Paper).

9.045 SANDHU, Meena (comp)
Automation and the silicon chip: a bibliography. London: G.L.C., 1979. 17 leaves. (G.L.C. Research Library bibliography No.106).

9.046 STUBBS, P.C. and METCALFE, J.S.
The economics of technological change and innovation: a select bibliography. Manchester: University of Manchester, Department of Economics, 1978. 61p.

9.047 TECHNOLOGY POLICY UNIT - University of Aston
Biotechnology: a review and annotated bibliography. Prepared by Harry Rothman, Richard Stanley, Susan Thompson and Zbigniew Towalski. London: Francis Pinter, 1981. 141p.

9.048 TRADES UNION CONGRESS LIBRARY
Technological change and employment: a bibliography. London: TUC, February, 1979. 5p. Mimeo.

9.049 UNITED STATES - Congress House Committee on Foreign Affairs: Subcommittee on National Security Policy and Scientific Developments
Science, technology, and American diplomacy: a selected, annotated bibliography of articles, books, documents, periodicals, and reference guides. Prepared by the Science Policy Research and Foreign Affairs Divisions, Legislative Reference Service, Library of Congress. Washington, USGPO, 1970.

9.050 UNITED STATES - National Aeronautics and Space Administration
Technology transfer - a selected bibliography. By M.Terry

Sovel. Prepared by University of Denver. Washington: USGPO, 1969.

9.051   UNITED STATES - National Science Foundation Office
        of National R & D Assessment
The effect of innovation on productivity in the service
industries.   3   volumes: Vol.1 Summary Report. Vol.2
Analytical paper. Vol.3 Annotated bibliography. Washington
D.C.: N.S.F., 1975. 15 leaves, 120p, not paginated.

9.052   UNIVERSITY OF UTAH - Human Resources Institute
Mechanisms for aiding worker adjustment to technological
change: concepts, review of the literature, abstracts.
(Vol.2 - key word index and abstracts). University of Utah,
no date. Not paginated.

9.053   WHISTON, Tom, SENKER, Peter and MACDONALD, Petrine
An annotated bibliography on the relationship between
technological change and educational development.   Paris:
UNESCO, International Institute for Educational Planning,
1980. 168p.

List Of Relevant Journals And Periodicals

American Economic Review
Economic Journal
Futures
Harvard Business Review
IEEE Transactions on Engineering Management
Journal of Political Economy
Long Range Planning
Nature
New Scientist
Omega
Patent Information Network (British Library)
R & D Management
R & D Management Digest
Research Management
Research Policy
Review of Economics and Statistics
Science Resources Newsletter (OECD)
Science, Technology and Human Values
Science and Technology Quarterly
Scientific American
Scientometrics
Sloan Management Review
Technological Forecasting and Social Change
Technology and Culture
Technology Review
Technology in Society
Technovation

Keyword Index

This index is arranged alphabetically according to keywords that describe the content of the references contained in this bibliography. It is intended to provide supplementary information only, as all references are contained within the chapters listed on the "Contents" page. Where users have not been able to find the information they seek through these chapter headings, the keyword index is intended as an alternative search mechanism.

References have been allocated up to four keywords and some may, therefore, appear several times in this index. Others, which were adequately described by the chapter headings, do not appear in the index at all.

Aerospace
   3.064   4.345   6.382   7.187   7.261

Agriculture
   4.082   4.083   6.016   6.275   9.031

Alternative Technology
   5.015   5.042

Argentina
   4.112   6.193   6.194   6.195

Australia
   1.052   2.058   3.038   3.184   4.003   5.002   6.019   6.147
   6.148   6.149   6.151   6.255   7.034   7.053   7.054   7.098
   7.124   7.125   7.126   7.127   7.128   7.129   7.159   7.323
   8.038

Austria
   3.125   7.101

Automation
   3.037   3.255   3.281   3.323   3.324   3.332   3.339   3.340
   3.360   3.363   4.021   4.232   4.241   4.293   5.052   5.053

```
       5.063   5.068   6.435   6.439   6.470   6.482   7.168   7.181
       7.213   7.256   7.318   8.002   8.003   8.006   8.009   8.010
       8.014   8.017   8.022   8.031   8.032   8.035   8.045   8.050
       8.051   8.052   8.053   8.056   8.059   8.061   8.063   8.068
       8.073   8.074   8.075   8.081   8.083   8.089   8.103   8.109
       8.120   8.123   8.126   8.128   8.132   8.141   9.045   9.052
```

Automobiles
```
       3.046   3.354   4.214   4.419   6.170   6.391   7.120   8.118
```

Banking
```
       3.332   3.372   8.012   8.030   8.035   8.057   8.135
```

Barriers
```
       3.008   7.002
```

Biotechnology
```
       2.060   2.074   3.032   3.137   3.257   3.311   3.312   3.313
       3.315   3.325   3.334   3.337   3.338   3.342   3.343   3.344
       3.348   3.352   3.353   3.355   3.361   3.362   3.364   3.365
       3.371   4.347   6.364   7.161   7.162   7.163   7.171   7.185
       7.186   7.212   7.234   7.249   7.277   7.283   7.296   7.351
       7.368   9.044   9.047
```

CAD
```
       3.307   3.314   3.322   3.351   3.356   4.290   7.172   8.006
```

CAD/CAM
```
       3.358   7.130   7.264   8.037
```

CAM
```
       3.339   3.340   3.359   7.260   7.265   8.037   8.129
```

Canada
```
       1.002   1.003   1.004   1.053   1.062   2.021   2.032   3.009
       3.017   3.023   3.173   3.199   3.204   3.220   3.225   3.226
       3.260   3.261   3.303   3.311   3.312   3.358   4.015   4.017
       4.020   4.072   4.073   4.076   4.225   4.226   4.245   4.257
       4.321   4.382   4.438   4.439   4.527   5.007   5.008   5.009
       6.048   6.049   6.151   6.457   7.012   7.036   7.054   7.081
       7.130   7.131   7.132   7.133   7.134   7.140   7.142   7.354
       8.025   8.144
```

Carpet Industry
```
       3.226   3.276   4.257
```

Chemicals
    1.063   3.001   3.003   3.005   3.130   3.201   3.257   3.301
    3.352   4.029   4.337   4.380   4.416   4.466   5.025   5.038
    6.064   6.140   6.167   6.176   6.219   6.339   6.484   7.093
    7.140   7.218   7.219   8.017

Clothing
    3.271   3.272   3.336   4.240   4.263   4.275   4.293   4.375
    4.463   4.511   7.207   8.046   8.050   8.056

Communications
    3.006   3.154   3.304   3.320   3.327   5.002   5.060   6.293
    7.064   7.195   7.344   7.349

Computer Industry
    4.226   4.243   4.393   4.395   4.472   7.252   8.064

Computer Systems
    3.216   3.217   4.074   4.372   5.071   6.481   7.175   7.179
    7.229   7.344   8.144

Computers
    3.284   3.285   4.234   4.296   7.071   7.221   7.311   8.047
    8.070   8.106   9.017

Costs
    3.046   3.060   3.132   6.142   6.302   6.441   6.499

Denmark
    4.011   4.019   5.004

Design
    3.040   3.108   3.109   3.159   3.161   3.194   3.269   4.358
    4.362   6.045   6.182   6.276   6.427   6.500   7.210   8.060
    8.061

Development
    2.034   3.205   3.234   4.004   4.020   4.022   4.078   4.371
    4.434   4.435   4.449   4.451   5.026   6.014   6.403   7.111
    7.136   7.361   9.008

Education
    6.420   7.176   7.304   8.024   8.028   8.036   8.048   8.058
    8.070   8.071   8.143   9.040   9.053

Electrical
    3.253   4.005   4.274   4.340

Electronics
    3.066   3.071   3.085   3.097   3.107   3.293   3.309   3.318
    3.321   3.329   3.330   3.333   3.341   3.345   3.346   3.349

3.350	3.357	3.367	3.368	3.373	4.055	4.215	4.227
4.281	4.297	4.312	4.322	4.331	4.332	4.369	4.403
4.427	4.475	4.492	4.512	5.072	6.051	6.085	6.468
6.492	7.070	7.192	7.193	7.196	7.199	7.209	7.214
7.220	7.345	7.352	7.357	8.115	8.131	9.005	9.010
9.018	9.023	9.026					

Employment
3.023	3.048	3.316	3.359	4.010	4.032	4.033	4.034
4.035	4.036	4.040	4.043	4.045	4.053	4.063	4.065
4.071	4.080	4.096	4.097	4.105	4.107	4.108	4.188
4.239	4.316	4.406	4.513	4.529	4.536	4.538	5.002
5.007	5.053	5.054	5.059	5.069	7.107	7.176	7.213
7.221	7.258	7.342	7.349	8.001	8.002	8.003	8.004
8.006	8.007	8.008	8.009	8.010	8.011	8.014	8.016
8.017	8.018	8.021	8.023	8.025	8.027	8.029	8.030
8.031	8.032	8.033	8.034	8.040	8.042	8.043	8.047
8.052	8.053	8.054	8.056	8.057	8.058	8.059	8.062
8.063	8.065	8.067	8.071	8.072	8.075	8.078	8.079
8.080	8.082	8.083	8.084	8.085	8.086	8.088	8.089
8.090	8.091	8.092	8.093	8.094	8.095	8.098	8.099
8.100	8.102	8.103	8.104	8.105	8.106	8.107	8.110
8.111	8.112	8.113	8.114	8.115	8.116	8.117	8.118
8.120	8.121	8.122	8.123	8.124	8.125	8.127	8.128
8.129	8.130	8.131	8.132	8.133	8.134	8.136	8.137
8.138	8.139	8.140	8.142	8.143	8.144	9.007	9.016
9.019	9.020	9.040	9.048				

Energy
1.045	1.048	3.145	3.263	3.270	3.290	3.291	3.310
4.109	4.518	5.046	6.070	6.145	6.345	6.397	7.056
7.164							

Engineering
2.010	3.040	3.082	3.155	3.224	3.277	3.322	4.352
4.356	4.415	4.510	4.520	6.113	6.132	6.223	6.228
6.274	6.277	6.300	6.401	6.427	6.507	7.181	7.202
7.228	8.006	8.014	8.026	8.036	8.041	8.045	8.066
8.069	8.074	8.075					

Europe
1.063	2.046	2.064	3.146	4.162	4.178	4.243	4.246
4.253	4.339	4.377	4.458	4.460	4.481	5.017	6.070
6.074	6.170	6.199	6.324	6.343	6.344	6.345	6.346
6.347	6.351	7.019	7.025	7.032	7.033	7.270	7.271
7.272	7.273	7.280	7.285	7.286	7.287	7.292	7.293
7.294	7.296	7.302	8.080	8.091	9.007		

F.R. Germany
2.062	3.351	4.029	4.044	4.219	4.336	4.414	4.418
4.479	4.480	6.091	6.466	7.052	7.154	7.155	7.156

7.157   7.325

Food
    4.244   6.218

Footwear
    3.036   3.103   4.317

Forecasting
    5.058   5.071   8.013

Foreign Investment
    4.235   4.239   4.341   4.379   4.487   4.496   7.323

France
    3.313   6.448   7.044   7.082   7.123

Global Systems
    4.030   4.057

Industrial Relations
    4.316   5.008   5.024   8.031

Informal Economy
    3.377   8.076   8.107   8.108

Information Technology
    3.305   3.306   3.326   3.331   3.366   3.370   4.228   5.009
    5.048   5.050   5.062   5.066   5.070   7.135   7.239   7.276
    8.005   8.013   8.015   8.020   8.080   8.104   8.124   8.135
    9.007

Information Transfer
    3.091   3.265   6.245   6.253   6.281   6.292   7.326

Instrumentation
    3.153   3.192   4.350   4.403   6.409   6.439   7.309

Intermediate Technology
    3.375   5.042   8.140

Ireland
    1.065   3.336   4.240   6.006   6.007   6.127   7.143   7.144
    7.145   7.146   7.147   7.148   7.149   7.150   7.151   7.152
    7.153   7.245   8.046   8.056   8.090

Iron and Steel
    3.308   7.117

Israel
    4.403   7.235

Japan
    3.110   4.074   4.088   4.101   4.123   4.128   4.141   4.169
    4.200   4.251   4.379   4.427   4.428   6.102   6.169   6.221
    6.435   6.463   7.070   7.158   7.180   7.182   7.232   7.233
    7.236   7.237   7.238   7.239

Latin America
    2.042   4.242   4.440   4.443   4.448   4.525

Leisure
    8.107   8.108

Licensing
    2.023   2.029   2.067   3.289   4.432   4.452   4.461   4.467
    4.484   4.488   4.489   4.507   6.493   7.235

Machine Tools
    3.019   3.085   3.281   4.252   4.353   7.137   7.318   8.067

Man-made Fibres
    4.374   7.197   7.310

Marketing
    6.174   6.205   6.218   6.258   6.272   6.278   6.431   6.461
    6.462   6.490   6.510

Materials
    3.310   4.067   4.189

Measuring Science & Technical Change
    1.006   1.010   1.011   1.037   1.038   1.055   1.060   1.065
    1.067   1.076   1.079   1.080   1.081   1.082   1.083   1.084
    1.085   1.086   1.087   1.088   1.089   1.090   1.091   2.024
    2.061   2.063   2.068   2.070   4.058   4.401   6.370   6.382
    6.457   9.043

Medical
    3.214   3.237   3.262   3.273   6.347

Microelectronics
    1.049   3.271   3.272   3.279   3.302   3.308   3.309   3.315
    3.316   3.317   3.318   3.319   3.320   3.321   3.329   3.330
    3.331   3.333   3.334   3.336   3.341   3.345   3.346   3.349
    3.350   3.354   3.357   3.360   3.367   3.368   3.369   3.373
    4.019   4.224   4.227   4.275   4.296   4.297   4.312   4.321
    4.331   4.332   4.354   4.376   4.536   4.538   5.004   5.049
    5.054   5.055   5.056   5.057   5.059   5.060   5.061   5.065

```
        5.066   5.067   5.072   6.411   6.468   7.101   7.176   7.209
        7.214   7.222   7.342   7.349   7.352   7.357   8.001   8.007
        8.008   8.012   8.013   8.016   8.021   8.024   8.025   8.028
        8.029   8.039   8.041   8.043   8.044   8.046   8.053   8.054
        8.055   8.071   8.072   8.077   8.079   8.082   8.090   8.091
        8.098   8.099   8.100   8.111   8.113   8.115   8.117   8.118
        8.119   8.125   8.130   8.133   8.134   8.135   8.136   8.137
        8.138   8.139   8.142   9.003   9.005   9.010   9.017   9.018
        9.019   9.020   9.021   9.023   9.024   9.025   9.026
```

Military Technology
```
        3.066   3.300   4.110   4.442   4.468   5.013   5.019   5.025
        6.009   6.091   6.162   6.312   6.491
```

Mining Machinery
```
        3.187   3.228
```

Multinationals
```
        3.152   4.253   4.270   4.292   4.316   4.338   4.343   4.369
        4.413   4.438   4.444   4.451   4.478   4.485   4.486   4.508
        4.519   6.026   6.256   7.327   7.332   9.008
```

Netherlands
```
        5.057   7.329
```

New Zealand
```
        3.182   3.183   6.206   7.075
```

Norway
```
        1.041   3.061   6.216   7.305
```

Nuclear Power
```
        3.240   3.270   5.031   5.033   5.035   5.037   6.398   7.045
```

Offices
```
        3.328   3.335   5.070   6.400   6.404   6.470   7.194   8.005
        8.008   8.009   8.010   8.015   8.023   8.034   8.053   8.081
        9.024   9.026
```

Operational Research
```
        6.012   6.060   6.402   6.440   6.442
```

Paper & Board
```
        3.016   4.224   7.191   8.019   8.065
```

Pharmaceuticals
```
        2.036   3.115   3.142   3.174   3.211   3.215   4.235   4.236
        4.237   4.282   4.349   4.399   4.400   4.461   5.043   6.039
        6.058   7.010   7.022   7.042   7.240   7.313   7.328
```

Plastics
  4.254   7.204   7.208   7.314   8.048   8.049   8.078

Politics of Technical Change
  5.001   5.015   5.027

Printing
  3.254   3.369   4.262   6.495   7.203   8.101

Public Policy
  3.008   3.262   4.062   4.132   4.138   4.155   4.311   4.321
  4.505   4.541   6.280   7.031   7.035   7.060   7.067   7.088
  7.090   7.093   7.095   7.096   7.241   7.242   7.243   7.259

Public Sector
  3.010   3.221   4.268

Regional
  4.528   4.529   4.530   4.531   4.532   4.533   4.534   4.535
  4.536   4.537   4.538   4.539   4.540   4.541   4.542   4.543
  4.544   4.545   9.009

Regulation
  3.060   3.078   3.132   3.240   5.038   5.043   5.044   5.045
  7.006   7.018   7.028   7.088   7.089   7.091   7.112   7.120
  7.173   7.263

Research Institutes
  2.052   6.081   6.229   6.233   6.234   6.235   6.236   6.424
  7.119   7.305   9.001

Research Policy
  4.334   4.352   6.063   6.076   6.204   6.217   6.218   6.350
  7.145   7.146   7.152   7.153   7.227   9.039

Resource Allocation
  5.019   6.014   6.015   6.257   6.309   6.312   9.041

Robotics
  3.230   3.363   5.051   5.068   6.476   7.168   8.026   8.038
  8.045   8.060   8.141   9.022

Rubber
  3.246   6.313

Science & Technology Links
  2.049   3.054   3.055   3.056   3.069   3.070   3.077   3.079
  3.107   3.118   3.143   3.144   3.162   3.178   3.179   4.170
  4.528   5.026   5.028   6.051   6.105   6.237   6.238   6.305
  6.432   6.449

Scandinavia
    4.111   7.059

Semiconductors
    3.058   3.106   3.294   4.205   4.246   4.248   4.404   4.417
    4.464   4.483   7.121   7.166   7.257

Services
    3.197   3.273   3.377   7.331   8.016   9.051

Skills
    3.023   5.068   8.007   8.010   8.018   8.019   8.029   8.030
    8.033   8.041   8.042   8.043   8.048   8.049   8.060   8.065
    8.066   8.067   8.069   8.073   8.074   8.078   8.085

Small Firms
    3.158   4.218   4.238   4.255   4.256   4.261   4.284   4.288
    4.355   4.359   4.360   4.361   4.387   4.396   4.407   4.408
    4.541   5.042   6.029   6.054   6.093   6.294   7.002   7.094
    7.122   7.183   7.334   9.015

Social Effects
    3.317   3.319   3.372   5.009   5.049   5.050   5.051   5.054
    5.055   5.056   5.057   5.058   5.059   5.060   5.061   5.062
    5.064   5.065   5.066   5.067   5.069   6.386   8.020   8.029
    8.060   9.003   9.016   9.018

Social Innovation
    3.374   3.375   3.376   3.378

Software
    4.354   8.039   8.104

Space Technology
    3.041   3.300

Steam Power
    4.006

Steel
    3.002   3.110   3.129   3.204   3.241   3.279   3.308   4.074
    4.376   6.040   6.193   6.194   6.195   6.199   8.030

Sweden
    3.013   3.109   3.224   3.334   4.023   4.024   4.025   4.026
    4.054   4.230   4.287   4.334   4.411   4.469   6.073   6.173
    6.217   6.218   6.300   7.030

Systems
    3.216   3.217   4.030   4.057   4.074   4.372   6.202   6.481
    7.175   7.179   7.229

Technology Choice
      4.010   4.047   4.351   4.451   5.011   5.012   5.042

Technology Transfer
      3.009   3.031   3.071   3.074   3.087   3.088   3.177   3.203
      3.206   3.212   3.213   3.218   3.232   3.247   3.249   3.297
      3.299   3.306   4.022   4.088   4.182   4.197   4.242   4.298
      4.306   4.307   4.379   4.388   4.396   4.424   4.430   4.433
      4.434   4.436   4.439   4.440   4.441   4.442   4.443   4.444
      4.445   4.447   4.448   4.449   4.456   4.457   4.458   4.464
      4.468   4.470   4.473   4.474   4.478   4.480   4.482   4.491
      4.492   4.493   4.494   4.499   4.500   4.504   4.505   4.508
      4.509   4.517   4.519   4.521   4.525   4.527   6.005   6.006
      6.007   6.009   6.067   6.082   6.090   6.127   6.150   6.248
      6.279   6.289   6.294   6.295   6.299   6.305   6.415   6.449
      6.456   6.463   6.491   6.502   7.003   7.013   7.024   7.028
      7.048   7.110   7.225   7.235   7.236   7.242   7.250   7.261
      7.317   7.323   7.324   7.327   7.328   7.332   7.333   7.338
      7.341   7.366   9.008   9.028   9.050

Telecommunications
      3.220   3.320   5.002   7.064   7.258   8.033   8.077   8.079

Textile Machinery
      3.155   3.156   3.157   4.232   6.264   6.265   6.267   6.268
      6.270   6.271   6.273   7.200   8.062

Textiles
      3.035   3.061   3.075   3.084   3.122   3.225   3.235   3.245
      3.282   4.375   4.377   4.378   4.436   4.463   4.511   6.262
      6.263   7.201   7.205   7.206   7.223

Tractors
      3.028   3.029   3.165   4.291

Training
      6.411   8.028   8.048

Transport
      3.101   3.163   6.249

U.K.
      3.015   3.028   3.029   3.047   3.058   3.112   3.173   3.224
      3.235   3.351   4.044   4.219   4.252   4.336   4.378   4.414
      4.446   6.084   7.118   8.020

U.S.A.
      1.021   1.022   1.023   1.024   1.025   1.026   1.027   1.028
      1.029   1.030   1.031   1.032   1.040   1.045   1.046   1.047
      1.048   1.049   1.050   1.051   1.060   1.061   1.075   1.076

```
      1.091   2.029   2.031   2.041   2.047   2.054   2.075   3.019
      3.036   3.058   3.076   3.133   3.173   3.204   3.235   3.296
      3.297   3.363   4.082   4.122   4.134   4.136   4.143   4.193
      4.206   4.207   4.220   4.235   4.236   4.237   4.265   4.270
      4.276   4.406   4.407   4.408   4.409   4.413   4.414   4.428
      4.429   4.437   4.460   4.471   4.473   4.475   4.480   4.481
      4.484   4.500   4.505   4.522   4.523   6.017   6.084   6.095
      6.155   6.186   6.190   6.199   6.329   6.330   6.338   6.385
      7.026   7.040   7.046   7.073   7.117   7.120   7.242   7.246
      7.247   7.248   7.249   7.251   7.252   7.253   7.254   7.255
      7.257   7.258   7.259   7.260   7.262   7.264   7.267   7.268
      7.269   8.027   8.031   9.011   9.049
```

Unemployment
```
      4.011   4.032   4.045   4.063   4.065   4.188   8.081   8.083
      8.084   8.093   8.103   8.105   8.107   8.108   8.110   8.112
      8.114   8.117   8.125   8.128   8.143
```

Unions
```
      5.022   7.033   8.001   8.008   8.009   8.011   8.012   8.047
      8.055   8.063   8.077   8.079   8.081   8.098   8.106   8.119
      8.137   8.138   8.139   8.142
```

Women
```
      1.020   5.029   8.004   8.007   8.015   8.043   8.057   8.072
      8.076   8.080   8.095   8.112   8.130
```

Work Organisation
```
      3.360   5.068   6.405   6.509   8.026   8.047   8.059   8.061
      8.109   9.025   9.052
```

Work Patterns
```
      8.087   9.017
```